M336
Mathematics and Computing: a third-level course

# GROUPS & GEOMETRY

## UNIT GR4
## FINITE GROUPS 1

Prepared for the course team by
**Bob Coates & Bob Margolis**

The Open University

This text forms part of an Open University third-level course.
The main printed materials for this course are as follows.

*Block 1*
Unit IB1   Tilings
Unit IB2   Groups: properties and examples
Unit IB3   Frieze patterns
Unit IB4   Groups: axioms and their consequences

*Block 2*
Unit GR1   Properties of the integers
Unit GR2   Abelian and cyclic groups
Unit GE1   Counting with groups
Unit GE2   Periodic and transitive tilings

*Block 3*
Unit GR3   Decomposition of Abelian groups
Unit GR4   Finite groups 1
Unit GE3   Two-dimensional lattices
Unit GE4   Wallpaper patterns

*Block 4*
Unit GR5   Sylow's theorems
Unit GR6   Finite groups 2
Unit GE5   Groups and solids in three dimensions
Unit GE6   Three-dimensional lattices and polyhedra

The course was produced by the following team:

Andrew Adamyk (BBC Producer)
David Asche (Author, Software and Video)
Jenny Chalmers (Publishing Editor)
Bob Coates (Author)
Sarah Crompton (Graphic Designer)
David Crowe (Author and Video)
Margaret Crowe (Course Manager)
Alison George (Graphic Artist)
Derek Goldrei (Groups Exercises and Assessment)
Fred Holroyd (Chair, Author, Video and Academic Editor)
Jack Koumi (BBC Producer)
Tim Lister (Geometry Exercises and Assessment)
Roger Lowry (Publishing Editor)
Bob Margolis (Author)
Roy Nelson (Author and Video)
Joe Rooney (Author and Video)
Peter Strain-Clark (Author and Video)
Pip Surgey (BBC Producer)

With valuable assistance from:

Maths Faculty Course Materials Production Unit
Christine Bestavachvili (Video Presenter)
Ian Brodie (Reader)
Andrew Brown (Reader)
Judith Daniels (Video Presenter)
Kathleen Gilmartin (Video Presenter)
Liz Scott (Reader)
Heidi Wilson (Reader)
Robin Wilson (Reader)

The external assessor was:
Norman Biggs (Professor of Mathematics, LSE)

The Open University, Walton Hall, Milton Keynes, MK7 6AA.

First published 1994. Reprinted 1997, 2002, 2003, 2007.

Copyright © 1994 The Open University

All rights reserved. No part of this publication may be reproduced, stored in a retrieval system or transmitted in any form or by any means, without written permission from the publisher or a licence from the Copyright Licensing Agency Limited. Details of such licences (for reprographic reproduction) may be obtained from the Copyright Licensing Agency Ltd of 90 Tottenham Court Road, London, W1P 9HE.

Edited, designed and typeset by the Open University using the Open University TEX System.

Printed in Malta by Gutenberg Press Limited.

ISBN 0 7492 2166 6

This text forms part of an Open University Third Level Course. If you would like a copy of *Studying with the Open University*, please write to the Central Enquiry Service, PO Box 200, The Open University, Walton Hall, Milton Keynes, MK7 6YZ. If you have not already enrolled on the Course and would like to buy this or other Open University material, please write to Open University Educational Enterprises Ltd, 12 Cofferidge Close, Stony Stratford, Milton Keynes MK11 1BY, United Kingdom.

# CONTENTS

| | |
|---|---|
| Study guide | 4 |
| Introduction | 5 |
| 1  Finite Abelian groups | 6 |
| 2  Subgroups of Abelian groups | 10 |
| 3  Permutation groups (audio-tape section) | 15 |
| 4  Conjugacy | 26 |
| 5  $p$-Groups | 32 |
| Solutions to the exercises | 37 |
| Objectives | 49 |
| Index | 50 |

# STUDY GUIDE

The sections in this unit are of roughly equal length as regards study time.

There is an audio programme associated with Section 3.

# INTRODUCTION

In *Units GR2* and *GR3*, our investigations were largely restricted to Abelian groups. In this and the next two Groups units we shall look at groups more generally, though we shall restrict ourselves to *finite* groups.

The most general problem that we might pose asks for a description of all possible finite groups. So, for each positive integer $n$, we might ask how many different groups, up to isomorphism, there are having that order and, for each of these groups, we might ask for some sort of canonical description. Unfortunately, this general problem is unsolved.

On the other hand it is possible to give a description of all finite groups in some cases where we impose some additional condition(s).

As an example, we already know from *Unit GR1* that, if we impose the condition that the order of the group is prime, then there is only one group of that order and it is cyclic.

> When we say 'only one', we mean 'only one up to isomorphism'. We shall not always include the phrase 'up to isomorphism', but it is always implied in such statements.

If the additional condition is that the group is Abelian, again a description of all Abelian groups of order $n$ is possible. We give such a description in Section 1, based on the work of *Unit GR3*.

If the additional condition is that the order of the group is a power of a prime, then it is possible to make *some* assertions about the structure of such groups. We consider this problem in Section 5.

As we saw in *Unit GR2*, some groups are isomorphic to a direct product of two or more of their subgroups. In such cases, knowledge of the group may be obtained from knowledge of the subgroups. Hence, another theme of this unit is investigating the existence of subgroups corresponding to divisors of the order of the group. We seek to generalize the very strong result for finite cyclic groups that we proved in *Unit GR2*, namely that a cyclic group has a *unique* subgroup corresponding to each and every divisor of its order. Moving from finite cyclic to finite Abelian groups produces a slightly weaker result. This result is discussed in Section 2.

From Section 3 onwards, we drop the condition that the groups considered are Abelian. The most general finite non-Abelian groups are, in a sense, the permutation groups $S_n$, for $n > 2$, and these are discussed in Section 3. This section also provides a counterexample to the converse of Lagrange's Theorem by exhibiting a group of order 12 which has no subgroup of order 6. It also reviews the idea of conjugacy in the context of the permutation groups $S_n$.

Conjugacy, for general groups, is discussed in Section 4. There we use the fact that a group action partitions the set on which it acts into distinct orbits to obtain a useful result known as the *class equation*.

In the final section of the unit, Section 5, we apply the class equation to obtain results about subgroups of groups of prime power order.

# 1 FINITE ABELIAN GROUPS

In this section we apply the Canonical Decomposition Theorem for Finitely Generated Abelian Groups, proved in Section 5 of *Unit GR3*, to the special case of *finite* Abelian groups.

Suppose that $A$ is a finite, non-trivial Abelian group of order $n$, that is

$$A = \{a_1, \ldots, a_n\}, \quad n > 1.$$

First we observe that $A$ is finitely generated because

$$A = \langle a_1, \ldots, a_n \rangle.$$

In other words, $A$ is generated by all its elements, although $A$ may well have a set of generators having considerably fewer than $n$ elements. (For example, we may omit the identity element $e$ from a set of generators, and the inclusion of an element $a$ in the set of generators guarantees that we need not also include any powers of $a$.)

Hence we can apply the Canonical Decomposition Theorem for Finitely Generated Abelian Groups to deduce that

$$A = \mathbb{Z}_{d_1} \times \cdots \times \mathbb{Z}_{d_k},$$

where

$$d_i \in \mathbb{Z}, \quad d_i > 1 \text{ or } d_i = 0, \quad d_i \mid d_{i+1}, \; i = 1, \ldots, k-1.$$

Because $A$ is a *finite* group, none of the terms in the direct product can be the infinite group $\mathbb{Z}$. This means that none of the $d_i$s can be zero. Thus we have

$$A = \mathbb{Z}_{d_1} \times \cdots \times \mathbb{Z}_{d_k},$$

where

$$d_i \in \mathbb{Z}, \quad d_i > 1, \quad d_i \mid d_{i+1}, \; i = 1, \ldots, k-1.$$

Furthermore, since $\mathbb{Z}_{d_i}$ has $d_i$ elements, this direct product has

$$d_1 \times \cdots \times d_k$$

elements and so

$$n = d_1 \times \cdots \times d_k.$$

We can view this application of the Canonical Decomposition Theorem and the above remarks from a slightly different viewpoint. If *all* we know about $A$ is its order, $n$, then each set of $d_i$s satisfying the conditions

$$d_i \in \mathbb{Z}, \quad d_i > 1, \quad i = 1, \ldots, k,$$
$$n = d_1 \times \cdots \times d_k,$$
$$d_i \mid d_{i+1}, \quad i = 1, \ldots, k-1,$$

determines the Abelian group

$$A = \mathbb{Z}_{d_1} \times \cdots \times \mathbb{Z}_{d_k},$$

of order $n$. Furthermore, because of the uniqueness of the torsion coefficients, different sets of such $d_i$s determine different Abelian groups. Thus, by finding all such sets of $d_i$s, we can describe precisely all the Abelian groups of a given order.

Finite Abelian groups have rank zero.

The problem of finding all such sets of $d_i$s is made easier by the divisibility property. Because of the divisibility property, any divisor of some $d_i$ must also divide all subsequent ones. We shall apply this observation to divisors which are powers of primes.

Because the order $n$ of $A$ is the product of the $d_i$s, any prime divisor $p$ of $n$ must appear as a prime divisor of at least one of the $d_i$s. Furthermore, by the divisibility property of the $d_i$s, the prime $p$ must divide all subsequent $d_i$s and, hence, certainly divides the *last* one, $d_k$. The following example indicates how this helps us determine all the Abelian groups of a given order.

**Example 1.1**

We show how to list all the Abelian groups of order 360.

We first need to find all sets of $d_i$s satisfying

$d_i \in \mathbb{Z}, \quad d_i > 1, \quad i = 1, \ldots, k,$
$360 = d_1 \times \cdots \times d_k,$
$d_i \mid d_{i+1}, \quad i = 1, \ldots, k - 1.$

Writing 360 as a product of primes, we have

$360 = 2^3 \times 3^2 \times 5^1.$

We shall usually omit exponents which are 1.

We consider the prime divisors of 360 starting with the largest, 5.
As 5 divides 360 and

$360 = d_1 \times \cdots \times d_k,$

it follows that 5 must divide one of the $d_i$s.
However, if it divides a $d_i$ *other than the last*, it would divide all subsequent ones and therefore

$360 = d_1 \times \cdots \times d_k$

would be divisible by $5^2$ or some higher power of 5.
Since the highest power of 5 dividing 360 is $5^1$, we are forced to the conclusion that 5 is only a factor of $d_k$, the last $d_i$.

Now we look at the prime power divisor $3^2$ of 360.
Arguing as above, if any $d_i$ has a factor $3^2$ (the highest power of 3 dividing 360), it must be the last one, $d_k$.
If no $d_i$ has a factor $3^2$, then the last *two* $d_i$s, that is $d_k$ and $d_{k-1}$, must each have a factor of 3.

Lastly, we look at the prime power divisor $2^3$. There are more possibilities here.
If $2^3$ divides any $d_i$, it must divide the last one, $d_k$.
If not, but $2^2$ divides some $d_i$, then again it must be the last one, $d_k$, and then $d_{k-1}$ must have a factor of 2.
The remaining possibility is that, of the last three $d_i$s, that is $d_k$, $d_{k-1}$ and $d_{k-2}$, each has a factor of 2.

We summarize the discussion above in the following table. This table lists the possible prime power divisors of the $d_i$s, working backwards from $d_k$.

| prime power | factors of $d_{k-2}$ | $d_{k-1}$ | $d_k$ | label |
|---|---|---|---|---|
| 5 | | | 5 | 5a |
| $3^2$ | | | $3^2$ | 3a |
| | | 3 | 3 | 3b |
| $2^3$ | | | $2^3$ | 2a |
| | | 2 | $2^2$ | 2b |
| | 2 | 2 | 2 | 2c |

For convenience in explaining the construction of the torsion coefficients, we have labelled the choices corresponding to each prime.

We now use this table to construct all the possibilities for the torsion coefficients $d_i$ and, hence, all Abelian groups of order 360.

We construct the torsion coefficients $d_i$ by observing that the possibilities for different prime powers are independent of one another. Thus, we may combine the only possibility for 5 with either of the two possibilities for $3^2$ and any of the three possibilities for $2^3$, giving

$1 \times 2 \times 3 = 6$

possibilities in all.

Without actually constructing the groups, we now know that there are six different Abelian groups of order 360.

We first construct $d_k$, the last torsion coefficient, working our way down the prime factors. The factors of $d_k$ must be 5, either $3^2$ or 3 and one of $2^3$, $2^2$ or 2. These possible values for $d_k$ are listed below, together with the rows of the table that produce them.

$$5 \times 3^2 \times 2^3 = 360 \quad \text{5a, 3a, 2a}$$
$$5 \times 3^2 \times 2^2 = 180 \quad \text{5a, 3a, 2b}$$
$$5 \times 3^2 \times 2 = 90 \quad \text{5a, 3a, 2c}$$
$$5 \times 3 \times 2^3 = 120 \quad \text{5a, 3b, 2a}$$
$$5 \times 3 \times 2^2 = 60 \quad \text{5a, 3b, 2b}$$
$$5 \times 3 \times 2 = 30 \quad \text{5a, 3b, 2c}$$

The torsion coefficient $d_k = 360$ corresponds to the choice 5a, 3a and 2a for the rows. These rows have no entries for $d_{k-1}$. Hence, there is only one torsion coefficient, that is $k = 1$, and

$$d_1 = 360,$$
$$A = \mathbb{Z}_{360}.$$

The value $d_k = 180$ corresponds to the choice 5a, 3a and 2b for the rows. The only entry for $d_{k-1}$, for this choice, is 2, so $d_{k-1} = 2$. There are no corresponding entries for $d_{k-2}$. Hence there are two torsion coefficients, that is $k = 2$, and

$$d_1 = 2,$$
$$d_2 = 180,$$
$$A = \mathbb{Z}_2 \times \mathbb{Z}_{180}.$$

For $d_k = 90$, corresponding to 5a, 3a and 2c, there are entries of 2 for both $d_{k-1}$ and $d_{k-2}$. Thus $d_{k-1} = 2$, $d_{k-2} = 2$ and there are three torsion coefficients, that is $k = 3$, and we have

$$d_1 = 2,$$
$$d_2 = 2,$$
$$d_3 = 90,$$
$$A = \mathbb{Z}_2 \times \mathbb{Z}_2 \times \mathbb{Z}_{90}.$$

A similar analysis can be done for the remaining choices and produces the following results.

Choice for $d_k$: 5a, 3b, 2a

$k = 2$

$d_1 = 3$

$d_2 = 5 \times 3 \times 2^3 = 120$

$A = \mathbb{Z}_3 \times \mathbb{Z}_{120}$

Choice for $d_k$: 5a, 3b, 2b

$k = 2$

$d_1 = 3 \times 2 = 6$

$d_2 = 5 \times 3 \times 2^2 = 60$

$A = \mathbb{Z}_6 \times \mathbb{Z}_{60}$

Choice for $d_k$: 5a, 3b, 2c

$k = 3$

$d_1 = 2$

$d_2 = 3 \times 2 = 6$

$d_3 = 5 \times 3 \times 2 = 30$

$A = \mathbb{Z}_2 \times \mathbb{Z}_6 \times \mathbb{Z}_{30}$

This discussion confirms our earlier observation that there are precisely six non-isomorphic Abelian groups of order 360. The groups are:

$$\mathbb{Z}_{2^3 \times 3^2 \times 5} = \mathbb{Z}_{360};$$
$$\mathbb{Z}_2 \times \mathbb{Z}_{2^2 \times 3^2 \times 5} = \mathbb{Z}_2 \times \mathbb{Z}_{180};$$
$$\mathbb{Z}_2 \times \mathbb{Z}_2 \times \mathbb{Z}_{2 \times 3^2 \times 5} = \mathbb{Z}_2 \times \mathbb{Z}_2 \times \mathbb{Z}_{90};$$
$$\mathbb{Z}_3 \times \mathbb{Z}_{2^3 \times 3 \times 5} = \mathbb{Z}_3 \times \mathbb{Z}_{120};$$
$$\mathbb{Z}_{2 \times 3} \times \mathbb{Z}_{2^2 \times 3 \times 5} = \mathbb{Z}_6 \times \mathbb{Z}_{60};$$
$$\mathbb{Z}_2 \times \mathbb{Z}_{2 \times 3} \times \mathbb{Z}_{2 \times 3 \times 5} = \mathbb{Z}_2 \times \mathbb{Z}_6 \times \mathbb{Z}_{30}.$$

The order of this listing corresponds to working from the highest prime downwards and, for each prime, working from the highest power downwards. ♦

*Exercise 1.1*

Using the approach in Example 1.1, find all Abelian groups of order:

(a) 900;

(b) 432.

In this exercise by the phrase 'find all Abelian groups' (of a given order) we mean write down all the ways of expressing groups of the given order as a direct product of cyclic groups corresponding to torsion coefficients, that is, in canonical form.

*Exercise 1.2*

Let $p$ be a prime. How many Abelian groups are there of order $p^5$?

*Exercise 1.3*

Let $p$, $q$ and $r$ be primes such that $p < q < r$. How many Abelian groups are there of each of the following orders?

(a) $p^3 q^2 r$

(b) $p^2 q^2 r^2$

(c) $p^4 q^3$

---

Once we have the prime decomposition of a positive integer $n$, the above technique enables us to find all Abelian groups of order $n$, displaying them in canonical form.

Finding the prime decomposition of a large integer is a non-trivial task.

Following Example 1.1 and the solutions to the above exercises, we can make the following observations.

Once we have obtained the prime decomposition of the order of the group:

(a) the number of columns in the table is the highest exponent appearing in the prime decomposition;

(b) the number of groups is the product of the number of choices (rows) corresponding to each prime divisor;

(c) having constructed the table, the possible sets of torsion coefficients and corresponding Abelian groups can be read off in a systematic manner.

By the remarks above, the *number* of Abelian groups of a given order $n$ depends only on the number of distinct primes in the factorization of $n$ and their exponents. The number of groups does not depend on the primes themselves.

For example, the analysis that showed that there are six Abelian groups of order

$$360 = 2^3 \times 3^2 \times 5$$

Example 1.1.

also shows that there are six Abelian groups of order

$$p^3 q^2 r,$$

Exercise 1.3.

for *any* choice of distinct primes $p$, $q$ and $r$. (In Exercise 1.3 we only imposed the condition $p < q < r$ to help you to construct the table in the same way as we did for 360.)

# 2 SUBGROUPS OF ABELIAN GROUPS

In Section 1 we considered the question of what we could say about the structure of a finite Abelian group just from a knowledge of its order. We were able to give a complete list of all such groups, expressing them in canonical form as direct products of cyclic groups corresponding to the torsion coefficients. Now we turn to the problem of what we can say about the existence and/or uniqueness of subgroups of a finite Abelian group.

To recap, for cyclic groups we know from *Unit GR2* that, if the cyclic group $A$ has order $n$ and if $m$ is a positive divisor of $n$, then $A$ has a unique subgroup of order $m$. We shall build on this result to investigate subgroups of Abelian groups.

> Cyclic groups possess both existence and *uniqueness* properties with regard to subgroups, for each positive divisor of the order of the group.

The next exercise shows that we cannot hope for such a strong result for Abelian groups in general.

*Exercise 2.1*

Find three elements of $\mathbb{Z}_2 \times \mathbb{Z}_2$ having order 2.

---

The result of Exercise 2.1 shows that, for finite Abelian groups, the answer to the subgroup problem is more complicated than for cyclic groups. Each of the three distinct elements of order 2 in $\mathbb{Z}_2 \times \mathbb{Z}_2$ gives rise to a subgroup of order 2 consisting of itself and the identity. Hence $\mathbb{Z}_2 \times \mathbb{Z}_2$ has *three* subgroups of order 2. So we know that in going from cyclic to Abelian groups we shall certainly have to sacrifice uniqueness of subgroups of a particular order.

*Exercise 2.2*

(a) What are the orders of the elements of $\mathbb{Z}_3 \times \mathbb{Z}_3$?

(b) How many subgroups of order 3 does $\mathbb{Z}_3 \times \mathbb{Z}_3$ possess?

---

For finite Abelian groups we have lost the uniqueness of the subgroup 'belonging' to a particular positive divisor of the order of the whole group. However, as we shall prove, subgroups still *exist* corresponding to every such divisor.

To establish this result we shall take the canonical decomposition as a direct product of cyclic groups and rearrange it in a more convenient form for this purpose. This rearrangement uses the fact we proved in *Unit GR2* that, if $m$ and $n$ are coprime, then

$$\mathbb{Z}_{mn} \cong \mathbb{Z}_m \times \mathbb{Z}_n.$$

We describe the process that we shall use in terms of a particular example.

**Example 2.1**

Suppose that $A$ is the Abelian group of order 432 whose canonical decomposition is

$$\mathbb{Z}_3 \times \mathbb{Z}_6 \times \mathbb{Z}_{24}.$$

We express each of the torsion coefficients 3, 6 and 24 as a product of prime powers.

$$3 = 3$$
$$6 = 2 \times 3$$
$$24 = 2^3 \times 3$$

Next we express each of the cyclic groups $\mathbb{Z}_3$, $\mathbb{Z}_6$ and $\mathbb{Z}_{24}$ as the corresponding direct product of cyclic groups of prime power order.

$$\mathbb{Z}_3 = \mathbb{Z}_3$$
$$\mathbb{Z}_6 \cong \mathbb{Z}_2 \times \mathbb{Z}_3$$
$$\mathbb{Z}_{24} \cong \mathbb{Z}_{2^3} \times \mathbb{Z}_3$$

Using these results, we express $A$ as a direct product of the cyclic groups of prime power order and then collect together terms corresponding to the same prime.

$$A \cong \mathbb{Z}_3 \times (\mathbb{Z}_2 \times \mathbb{Z}_3) \times (\mathbb{Z}_{2^3} \times \mathbb{Z}_3)$$
$$\cong (\mathbb{Z}_2 \times \mathbb{Z}_{2^3}) \times (\mathbb{Z}_3 \times \mathbb{Z}_3 \times \mathbb{Z}_3)$$

We have now expressed $A$ as (being isomorphic to) the product of two groups of prime power orders. The first is

$$\mathbb{Z}_2 \times \mathbb{Z}_{2^3},$$

of order $2^4 = 16$; the second is

$$\mathbb{Z}_3 \times \mathbb{Z}_3 \times \mathbb{Z}_3$$

of order $3^3 = 27$. These prime powers correspond precisely to the prime decomposition

$$432 = 2^4 \times 3^3.$$

The two groups in this new direct product are of coprime order. ♦

This process generalizes, and we may express any finite Abelian group as a direct product of groups each of which has prime power order. Since the components of this direct product correspond to different primes, their orders are coprime.

This form of decomposition of a finite Abelian group $A$ is called the **primary decomposition** of $A$, and the term in the direct product corresponding to a prime $p$ is called the ***p*-primary component** of $A$. Thus, in Example 2.1, the 2-primary component of $A$ is $\mathbb{Z}_2 \times \mathbb{Z}_{2^3}$ and the 3-primary component is $\mathbb{Z}_3 \times \mathbb{Z}_3 \times \mathbb{Z}_3$. Note that, in this case, the $p$-primary components are not cyclic, whereas the components in the canonical decomposition always are. So, in going from the canonical to the primary decomposition, we have gained the fact that the components have coprime orders, actually prime power orders, but lost the fact that the components must be cyclic.

### Exercise 2.3

Find the primary decomposition of

$$A = \mathbb{Z}_2 \times \mathbb{Z}_6 \times \mathbb{Z}_{30} \times \mathbb{Z}_{150}.$$

Hence write down the $p$-primary components of $A$ for each prime dividing the order of the group.

As we may deduce from the construction of primary decompositions in Example 2.1 and Exercise 2.3, the canonical decomposition of a finite Abelian group *uniquely* determines its primary decomposition. Furthermore, we can determine the canonical decomposition from a given primary decomposition by using the prime powers in the primary decomposition in a manner similar to that used in Section 1. For example, for the primary decomposition

$$A \cong (\mathbb{Z}_2 \times \mathbb{Z}_{2^3}) \times (\mathbb{Z}_3 \times \mathbb{Z}_3 \times \mathbb{Z}_3)$$

in Example 2.1, we take the highest prime powers, $2^3$ and 3, of each prime to give us

$$\mathbb{Z}_{2^3} \times \mathbb{Z}_3,$$

the next highest, 2 and 3, of each prime to give us

$$\mathbb{Z}_2 \times \mathbb{Z}_3,$$

and lastly the remaining prime power 3, to give us

$$\mathbb{Z}_3,$$

and combining these we get the canonical decomposition

$$A \cong \mathbb{Z}_3 \times (\mathbb{Z}_2 \times \mathbb{Z}_3) \times (\mathbb{Z}_{2^3} \times \mathbb{Z}_3)$$
$$\cong \mathbb{Z}_3 \times \mathbb{Z}_6 \times \mathbb{Z}_{24}.$$

Thus, because of the uniqueness of the canonical decomposition, two finite Abelian groups are isomorphic if and only if they have the same primary decomposition. Formalizing this statement, we have the following theorem.

### Theorem 2.1

Let $A$ and $B$ be two finite Abelian groups.
Then $A$ and $B$ are isomorphic if and only if the $p$-primary components of $A$ and $B$ are the same for every prime $p$.

We are now in a position to prove the following theorem.

### Theorem 2.2 Subgroups of finite Abelian groups

Let $A$ be a finite Abelian group of order $n$ and let $m$ be a positive divisor of $n$.
Then there exists a subgroup of $A$ of order $m$.

Before giving the proof, which necessarily uses quite involved notation, we illustrate the proof strategy with an example.

### Example 2.2

Let $A$ be the Abelian group of order 360 with canonical decomposition

$$A = \mathbb{Z}_6 \times \mathbb{Z}_{60}.$$

We shall show that $A$ has a subgroup of order 12.

We first obtain the primary decomposition of $A$:

$$A = \mathbb{Z}_6 \times \mathbb{Z}_{60}$$
$$\cong (\mathbb{Z}_2 \times \mathbb{Z}_3) \times (\mathbb{Z}_4 \times \mathbb{Z}_3 \times \mathbb{Z}_5)$$
$$\cong (\mathbb{Z}_2 \times \mathbb{Z}_4) \times (\mathbb{Z}_3 \times \mathbb{Z}_3) \times \mathbb{Z}_5$$
$$\cong A_1 \times A_2 \times A_3,$$

where $A_1$ is the 2-primary component $\mathbb{Z}_2 \times \mathbb{Z}_4$, $A_2$ is the 3-primary component $\mathbb{Z}_3 \times \mathbb{Z}_3$ and $A_3$ is the 5-primary component $\mathbb{Z}_5$.

We write 12, the order of the required subgroup, as a product of prime powers

$$12 = 2^2 \times 3.$$

We now construct a primary decomposition of a subgroup of order 12 by selecting subgroups of the primary components of $A$.

From the 2-primary component $A_1 = \mathbb{Z}_2 \times \mathbb{Z}_4$ we select a subgroup of order $2^2 = 4$. Perhaps the obvious one is

$$\{0\} \times \mathbb{Z}_4 \cong \mathbb{Z}_4.$$

On the other hand, because 2 divides 4 and $\mathbb{Z}_4$ is cyclic, $\mathbb{Z}_4$ has a cyclic subgroup of order 2 ($\{0, 2\} \cong \mathbb{Z}_2$). So we could have chosen

$$\mathbb{Z}_2 \times \{0, 2\} \cong \mathbb{Z}_2 \times \mathbb{Z}_2.$$

In either case, we can be sure that $A_1$ has a subgroup $B_1$ of order 4.

From $A_2 = \mathbb{Z}_3 \times \mathbb{Z}_3$ we select a subgroup of order 3. From Exercise 2.2 you have seen that there are a number of possible choices. As $B_2$ we choose any one of them (which must be isomorphic to $\mathbb{Z}_3$).

From $A_3$, we select the trivial subgroup of order 1 as $B_3$.

Finally, since $B_1$ is a subgroup of $A_1$, $B_2$ is a subgroup of $A_2$ and $B_3$ is a subgroup of $A_3$, we have that

$$B_1 \times B_2 \times B_3$$

is a subgroup of $A_1 \times A_2 \times A_3$ of order 12. Since

$$A \cong A_1 \times A_2 \times A_3,$$

there is a subgroup $B$ of $A$ such that

$$B \cong B_1 \times B_2 \times B_3,$$

and $B$ is a required subgroup of order 12. ♦

As you may check if you wish, the direct product of subgroups of a set of groups is always a subgroup of the direct product of the groups.

There are several observations that we can make based on the above example.

First, since $B_3$ is trivial, it may be omitted from the direct product, so that we have

$$B \cong B_1 \times B_2.$$

We shall generally omit trivial terms from now on.

Second, from now on we shall generally refer to $B_1 \times B_2$ as *being* a required subgroup of $A$, rather than being *isomorphic to* a required subgroup, and we shall generally write

$$B = B_1 \times B_2.$$

Third, the two different choices which were available for $B_1$ give rise to two non-isomorphic subgroups of $A$ of order 12, namely

$$B = \mathbb{Z}_4 \times \mathbb{Z}_3 \cong \mathbb{Z}_{12}$$

and

$$C = \mathbb{Z}_2 \times \mathbb{Z}_2 \times \mathbb{Z}_3 \cong \mathbb{Z}_2 \times \mathbb{Z}_6.$$

The choices available for $B_2$ are all the same (up to isomorphism) and so do not count as different.

Therefore the example shows that, in moving from finite cyclic to finite Abelian groups, not only have we lost the uniqueness of the subgroup corresponding to a particular divisor, but different subgroups of the same order need not even be isomorphic.

We now give a formal proof of Theorem 2.2. The strategy is to construct the primary decomposition corresponding to a subgroup of the required order by taking subgroups of the $p$-primary components of $A$.

*Proof of Theorem 2.2*

Suppose that the prime decomposition of $n$ is

$$n = p_1^{k_1} \ldots p_r^{k_r}, \quad 0 < k_i, i = 1, \ldots, r.$$

It follows that, as $m$ divides $n$, the prime decomposition of $m$ must be of the form

$$m = p_1^{l_1} \ldots p_r^{l_r}, \quad 0 \leq l_i \leq k_i, i = 1, \ldots, r.$$

Now, the primary decomposition of $A$ is of the form

$$A_1 \times \cdots \times A_r,$$

where each $A_i$ has order $p_i^{k_i}$.

For each $A_i$ we are going to find a subgroup $B_i \subseteq A_i$ such that $B_i$ has order $p_i^{l_i}$. Because each $B_i$ is a subgroup of the corresponding $A_i$, the direct product of the $B_i$s will be a direct product of subgroups of the $A_i$s; that is, it will be isomorphic to a subgroup $B$ of $A$. Furthermore,

$$B \cong B_1 \times \cdots \times B_r$$

will have the required order.

Each $p_i$-primary component $A_i$ is a direct product of cyclic groups, each of order a power of $p_i$, the sum of the powers being $k_i$. This corresponds to writing

$$p_i^{k_i} = p_i^{\alpha_1} \times \cdots \times p_i^{\alpha_s},$$

where

$$\alpha_1 + \cdots + \alpha_s = k_i.$$

Because $l_i \leq k_i$, we may write

$$l_i = \beta_1 + \cdots + \beta_s,$$

where

$$0 \leq \beta_j \leq \alpha_j, \quad j = 1, \ldots, s.$$

Because $\beta_j \leq \alpha_j$, $j = 1, \ldots, s$,

$$p_i^{\beta_j} \mid p_i^{\alpha_j}, \quad j = 1, \ldots, s,$$

and each $p_i^{\alpha_j}$ is the order of one of the *cyclic* components of the $p_i$-primary term $A_i$.

Hence the cyclic component of $A_i$ of order $p_i^{\alpha_j}$ has a subgroup of order $p_i^{\beta_j}$ for $j = 1, \ldots, s$.

Therefore the direct product, $B_i$, of all of these subgroups has order

$$p_i^{\beta_1 + \cdots + \beta_s} = p_i^{l_i}.$$

The above argument holds for each $A_i$. Therefore the subgroup

$$B \cong B_1 \times \cdots \times B_r$$

has order

$$m = p_1^{l_1} \ldots p_r^{l_r}, \quad 0 \leq l_i \leq k_i.$$

This completes the proof. ∎

> Each $A_i$ is the direct product of the cyclic groups corresponding to $p_i$ obtained by splitting up the components of the canonical decomposition, as we did in Examples 2.1 and 2.2.

> As we saw in Example 2.2, it may be possible to write $l_i$ as such a sum in several different ways. Each way gives rise to a subgroup of the required order.

So far we have shown that, for finite Abelian groups, we have retained the existence of subgroups corresponding to each positive divisor of the order but, unlike the finite cyclic case, they are no longer unique or even isomorphic.

> The loss of uniqueness and isomorphism was illustrated in Example 2.2 and in the third observation that followed it.

*Exercise 2.4*

Let $A$ be the Abelian group

$$\mathbb{Z}_2 \times \mathbb{Z}_2 \times \mathbb{Z}_8,$$

of order 32.

(a) Find two non-isomorphic subgroups of $A$ having order 4.

(b) Find three non-isomorphic subgroups of $A$ having order 8.

*Exercise 2.5*

Let $A$ be the Abelian group with primary decomposition

$$\mathbb{Z}_3 \times \mathbb{Z}_9 \times \mathbb{Z}_{25},$$

of order 675. Show that $A$ has two non-isomorphic subgroups of order 45, one of which is cyclic.

# 3 PERMUTATION GROUPS (AUDIO-TAPE SECTION)

In this section we shall remind you of some of the basic properties of the permutation groups $S_n$. We shall also show why permutation groups are, in some sense, the most general finite groups available.

The audio programme discusses conjugacy in $S_n$ and, as a result of this discussion, we shall show that the converse of Lagrange's Theorem is false.

*You should now listen to the audio programme for this unit, referring to the tape frames below when asked to during the programme.*

## 1 Outline

(a) Investigate conjugacy in permutation groups

(b) Produce a group of order 12 with no subgroup of order 6

*(converse of Lagrange's Theorem is false)*

## 2 Notation

$S_n$: set of permutations of set $\{1, 2, \ldots, n\}$

*(example in $S_8$)*

$$\begin{pmatrix} 1 & 2 & 3 & 4 & 5 & 6 & 7 & 8 \\ 2 & 5 & 1 & 7 & 3 & 4 & 6 & 8 \end{pmatrix}$$

$1 \mapsto 2,\ 2 \mapsto 5,\ \ldots,\ 7 \mapsto 6,\ 8 \mapsto 8$

## 3 Cycle notation

Notation (1 2 5 3) means

$1 \mapsto 2,\ 2 \mapsto 5,\ 5 \mapsto 3,\ 3 \mapsto 1$

*or*

$2 \mapsto 5,\ 5 \mapsto 3,\ 3 \mapsto 1,\ 1 \mapsto 2$

(2 5 3 1)

## 4 Cycle form

$$\begin{pmatrix} 1 & 2 & 3 & 4 & 5 & 6 & 7 & 8 \\ 2 & 5 & 1 & 7 & 3 & 4 & 6 & 8 \end{pmatrix}$$

(1 2 5 3) (4 7 6) (8)
or for example
(8) (4 7 6) (1 2 5 3)

*(disjoint cycles)*

### Exercise 3.1

Express the following elements of $S_8$ in cycle form, i.e. as products of disjoint cycles

(a) $\begin{pmatrix} 1 & 2 & 3 & 4 & 5 & 6 & 7 & 8 \\ 3 & 1 & 4 & 6 & 8 & 2 & 5 & 7 \end{pmatrix}$

(b) $\begin{pmatrix} 1 & 2 & 3 & 4 & 5 & 6 & 7 & 8 \\ 1 & 8 & 4 & 2 & 5 & 7 & 6 & 3 \end{pmatrix}$

## 4A

**Solution 3.1**

(a) (1 3 4 6 2) (5 8 7)

(b) (2 8 3 4) (6 7)

or for example

(a) (5 8 7) (1 3 4 6 2)  *(disjoint cycles commute)*

(b) (3 4 2 8) (7 6)  *(representation not unique)*

etc.

## 5

**Cycle types in $S_n$**

All representations of

(1 3 4 6 2) (5 8 7)

as a product of disjoint cycles have:

**one 5-cycle and one 3-cycle**

This is the **cycle type**

## 6

**Example**

Cycle types in $S_3$

*(3 = 1 + 1 + 1; 3 = 2 + 1; 3 = 3)*

| (*) (*) (*) | (* *) (*) | (* * *) |
|---|---|---|
| $e$ | (1 2) | (1 2 3) |
|  | (1 3) | (1 3 2) |
|  | (2 3) |  |
| 1 | 3 | 2 | = 6 |

## 7

**Exercise 3.2**

Cycle types in $S_4$

*(4 = 1 + 1 + 1 + 1; 4 = 2 + 1 + 1; etc.)*

| (*) (*) (*) (*) | (* *) (*) (*) | (* * *) (*) | (* * * *) | (* *) (* *) |
|---|---|---|---|---|
|  |  |  |  |  |
|  |  |  |  | = 24 |

## 7A

**Solution 3.2**

Cycle types in $S_4$

| (*)(*)(*)(*) | (**)(*)(*) | (***)(*) | (****) | (**)(**) |
|---|---|---|---|---|
| $e$ | (1 2) | (1 2 3) | (1 2 3 4) | (1 2)(3 4) |
| | (1 3) | (1 2 4) | (1 2 4 3) | (1 3)(2 4) |
| | (1 4) | (1 3 4) | (1 3 2 4) | (1 4)(2 3) |
| | (2 3) | (1 3 2) | (1 3 4 2) | |
| | (2 4) | (1 4 2) | (1 4 2 3) | |
| | (3 4) | (1 4 3) | (1 4 3 2) | |
| | | (2 3 4) | | |
| | | (2 4 3) | | |
| 1 | 6 | 8 | 6 | 3 | = 24

## 8

**Counting cycle types in $S_4$**

3-cycles

$$\frac{(4 \times 3 \times 2)}{3} \times 1$$

⟵ for representations

So eight 3-cycles in $S_4$

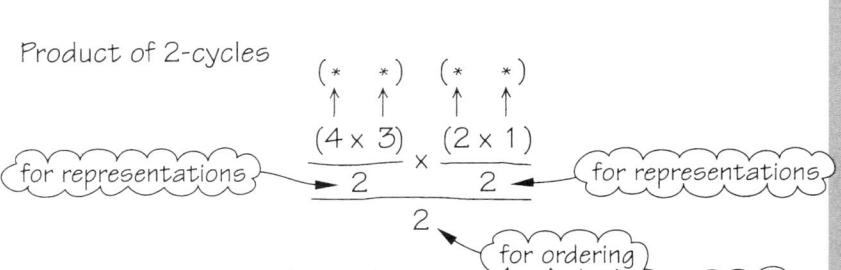

Product of 2-cycles

$$\frac{(4 \times 3)}{2} \times \frac{(2 \times 1)}{2}$$
$$\frac{}{2}$$

for representations ⟶  ⟵ for representations

for ordering
$(ab)(cd) = (cd)(ab)$

So three products of 2-cycles

## 9

**Exercise 3.3**

In $S_7$, how many permutations are there of each of the following types?

(a) 4-cycle and 2-cycle
   $(* * * *)(* *)(*)$

(b) two 3-cycles
   $(* * *)(* * *)(*)$

## 9A

**Solution 3.3**

(a) $\dfrac{7 \times 6 \times 5 \times 4}{4} \times \dfrac{3 \times 2}{2} \times 1 = 630$

(b) $\dfrac{\dfrac{7 \times 6 \times 5}{3} \times \dfrac{4 \times 3 \times 2}{3}}{2} \times 1 = 280$

## 10

**Cycle type gives conjugacy in $S_n$**

In $S_5$: $\quad x = (1\ 4\ 2)(3\ 5)$
$\quad\quad\quad y = (2\ 3\ 4)(1\ 5)$ ⟵ same cycle type

*Want* permutation $g$ such that $y = gxg^{-1}$

Define $g$

$x = (1 \xrightarrow{x} 4 \xrightarrow{x} 2)(3 \xrightarrow{x} 5)$
$\quad\ \ \downarrow g\ \ \downarrow g\ \ \downarrow g\ \ \downarrow g\ \ \downarrow g \quad\quad g = (1\ 2\ 4\ 3)$
$y = (2 \xrightarrow{y} 3 \xrightarrow{y} 4)(1 \xrightarrow{y} 5)$

$y: 2 \mapsto 3 \quad\quad\begin{array}{c} 1 \xrightarrow{x} 4 \\ g^{-1}\uparrow\ \downarrow g\quad\ \downarrow g \\ \boxed{2 \xrightarrow{y} 3} \end{array}$ ⟵ $g^{-1}$ first in $gxg^{-1}$

$\quad\quad\quad y = gxg^{-1}$

$x$ and $y$, of same cycle type, are conjugate

## 11

**Generally, same cycle types are conjugate**

If $x$ and $y$ have the same cycle type,
match up cycles of equal lengths

$x = (*\ *\ \cdots\ *)(*\ *\ \cdots\ *)\cdots(*)(*)$
$\quad\ \ g\downarrow\downarrow\ \ \ \ \ \downarrow\ \ \downarrow\downarrow\ \ \ \ \ \downarrow\ \ \ \ \ \ \ \downarrow\ \ \downarrow g$
$y = (*\ *\ \cdots\ *)(*\ *\ \cdots\ *)\cdots(*)(*)$

and define $g$, as shown, mapping the symbols of $x$ to those of $y$

$g$ conjugates $x$ to $y$: $\quad\quad i \xrightarrow{x} j$
$\quad\quad\quad\quad\quad\quad\quad g^{-1}\uparrow\ \downarrow g\quad\ \downarrow g$
$\quad\quad\quad\quad\quad\quad\quad g(i) \xrightarrow{y} g(j) \quad\quad y = gxg^{-1}$

## 12 — Conjugacy gives cycle type in $S_n$

$g$ and $x = (1\ 5\ 2)(3\ 4)$ in $S_5$

$$1 \xrightarrow{x} 5$$
$$g^{-1} \uparrow \qquad \downarrow g$$
$$g(1) \quad g(5) \qquad gxg^{-1}: g(1) \mapsto g(5)$$

$x: 1 \mapsto 5$

$gxg^{-1} = (g(1)\ g(5)\ g(2))(g(3)\ g(4))$

$x$ and its conjugate $gxg^{-1}$ have the same cycle type

True generally: the argument in Frame 11 works in reverse

## 13 — Calculating parity (odd or even)

$$x = \begin{pmatrix} 1 & 2 & 3 & 4 & 5 & 6 & 7 & 8 \\ 2 & 5 & 1 & 7 & 3 & 4 & 6 & 8 \end{pmatrix}$$

Crossing diagram

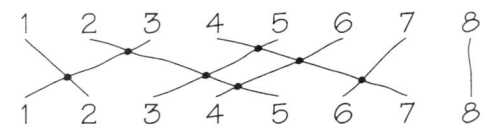

7 crossings: $x$ is odd

$x = (1\ 2\ 5\ 3)(4\ 7\ 6)(8)$

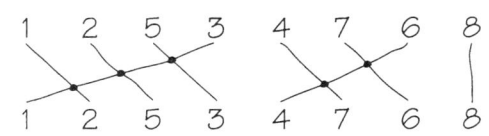

$3 + 2 = 5$ crossings: $x$ is odd

Cycles of odd length are even
Cycles of even length are odd

## 14 — Alternating group $A_n$

Homomorphism
$\emptyset: S_n \to \mathbb{Z}_2$
$\emptyset(\text{even}) = 0$
$\emptyset(\text{odd}) = 1$

*$A_n$ is the normal subgroup of $S_n$ consisting of even permutations*

$\text{Ker}(\emptyset) = A_n$, the **alternating group of degree $n$**

## 15 — Exercise 3.4

Determine the parity of the elements of $S_4$, and hence determine $A_4$

*Hint* Use Solution 3.2 in Frame 7A

## 15A

### Solution 3.4

Even permutations have the following cycle types:
$(*)(*)(*)(*)$
$(***)(*)$
$(**)(**)$

The other cycle types are odd

So
$A_4 = \{e, (1\,2\,3), (1\,2\,4), (1\,3\,4), (1\,3\,2), (1\,4\,2), (1\,4\,3),$
$\quad\quad (2\,3\,4), (2\,4\,3), (1\,2)(3\,4), (1\,3)(2\,4), (1\,4)(2\,3)\}$

## 16

### The group $A_4$

$|A_4| = 12$

### Exercise 3.5

Show $H = \{e, (1\,2)(3\,4), (1\,3)(2\,4), (1\,4)(2\,3)\}$
is a subgroup of $A_4$

## 16A

### Solution 3.5

*(closure ✓)*

$(1\,2)(3\,4)\,(1\,3)(2\,4) = (1\,3)(2\,4)\,(1\,2)(3\,4) = (1\,4)(2\,3)$
$(1\,2)(3\,4)\,(1\,4)(2\,3) = (1\,4)(2\,3)\,(1\,2)(3\,4) = (1\,3)(2\,4)$
$(1\,3)(2\,4)\,(1\,4)(2\,3) = (1\,4)(2\,3)\,(1\,3)(2\,4) = (1\,2)(3\,4)$

Every element is self-inverse   *(order 2)*

## 17

### $A_4$ has no subgroup $H$ of order 6

If so:

(a) $H$ would be isomorphic to $S_3$   *(no element of order 6)*

(b) $H$ would contain the only three elements of order 2 in $A_4$

(c) $H$ would have a subgroup with 4 elements
   (the one in Exercise 3.5)

CONTRADICTION   *(4 does not divide 6)*

Hence, the converse of Lagrange's Theorem is not true

The following exercise is designed to give you extra practice with some of the computational techniques from the tape.

*Exercise 3.6*

Let $x$ and $y$ be the following permutations in $S_6$:
$$x = \begin{pmatrix} 1 & 2 & 3 & 4 & 5 & 6 \\ 3 & 5 & 4 & 6 & 2 & 1 \end{pmatrix};$$
$$y = \begin{pmatrix} 1 & 2 & 3 & 4 & 5 & 6 \\ 6 & 3 & 4 & 5 & 2 & 1 \end{pmatrix}.$$

Explain why $x$ and $y$ are conjugate in $S_6$ and find a permutation $g \in S_6$ such that
$$y = gxg^{-1}.$$

Now we turn to our assertion that permutation groups are, in some sense, the most general finite groups available. By this we mean that every finite group is isomorphic to a subgroup of some permutation group.

To be specific, we shall show that the following theorem holds.

***Theorem 3.1 Cayley's theorem***

Let $G$ be a finite group of order $n$.
Then $G$ is isomorphic to a subgroup of the permutation group $S_n$.

The proof is constructive. That is, we shall show how to write down the permutation in $S_n$ corresponding to each element of $G$. Formally, we shall define a homomorphism $\phi$ from $G$ to $S_n$ and show that $\phi$ is one–one. Because $\phi$ is one–one, $G$ is isomorphic to the subgroup $\text{Im}(\phi)$ of $S_n$.

We first illustrate the definition of $\phi$ using an example.

**Example 3.1**

Let $G$ be the Klein group. We are going to define a homomorphism
$$\phi : G \to S_4.$$
The group $S_4$ is the set of all permutations of the set
$$\{1, 2, 3, 4\}.$$
To define $\phi$ we label the elements of $G$:
$$G = \{g_1, g_2, g_3, g_4\},$$
where $g_1$ is the identity of $G$. The Cayley table for $G$ is as follows:

|       | $g_1$ | $g_2$ | $g_3$ | $g_4$ |
|-------|-------|-------|-------|-------|
| $g_1$ | $g_1$ | $g_2$ | $g_3$ | $g_4$ |
| $g_2$ | $g_2$ | $g_1$ | $g_4$ | $g_3$ |
| $g_3$ | $g_3$ | $g_4$ | $g_1$ | $g_2$ |
| $g_4$ | $g_4$ | $g_3$ | $g_2$ | $g_1$ |

Each row of a Cayley table is created by left multiplying the column headings by the corresponding row heading. Furthermore, each row contains precisely the elements of the whole group. Hence, each row, given by left multiplication, defines a permutation of the elements of the group.

For example, left multiplication by $g_2$ gives rise to the row
$$g_2 \quad g_1 \quad g_4 \quad g_3.$$

The corresponding permutation of the elements of $G$ is

$$\begin{pmatrix} g_1 & g_2 & g_3 & g_4 \\ g_2 & g_1 & g_4 & g_3 \end{pmatrix}.$$

This, in turn, defines a corresponding element

$$\begin{pmatrix} 1 & 2 & 3 & 4 \\ 2 & 1 & 4 & 3 \end{pmatrix} = (12)(34)$$

of $S_4$. We define $\phi(g_2)$ to be this element of $S_4$.

Repeating this process for the remaining elements of $G$ gives the following definition of $\phi$:

$\phi(g_1) = e$;
$\phi(g_2) = (12)(34)$;
$\phi(g_3) = (13)(24)$;
$\phi(g_4) = (14)(23)$.

Effectively, what we have done is to read off the permutations by looking at the suffices on the elements in the rows of the Cayley table.

In this example, direct calculation shows that

$$\{e, (12)(34), (13)(24), (14)(23)\}$$

is a subgroup of $S_4$, which is certainly the Klein group because each non-identity element has order 2. Thus, $\text{Im}(\phi)$ is a subgroup of $S_4$ which is isomorphic to $G$. ♦

### Exercise 3.7

Let $G$ be the cyclic group

$$C_3 = \{a^0, a^1, a^2\}.$$

Rename the elements $g_1 = a^0$, $g_2 = a^1$ and $g_3 = a^2$ and use the technique of Example 3.1 to define the corresponding mapping $\phi$ from $G$ to $S_3$. Show that $\text{Im}(\phi)$ is a subgroup of $S_3$ which is isomorphic to $G$.

---

Although, in Example 3.1 and Exercise 3.7, we showed that $G \cong \text{Im}(\phi)$, we did *not* show that $\phi$ is a one–one homomorphism (i.e. that $\phi$ is an isomorphism from $G$ to $\text{Im}(\phi)$). We prove this in general as part of our proof of Cayley's Theorem, which follows. The proof involves generalizing the construction used in Example 3.1 and Exercise 3.7.

### Proof of Theorem 3.1

We must define the mapping $\phi$, show that it is one–one and show that it has the morphism property. Once we have done this, we know that $\text{Im}(\phi)$ is a subgroup of $S_n$ and hence that $\phi$ is an isomorphism from $G$ to $\text{Im}(\phi)$.

Let the $n$ elements of $G$ be $g_1, \ldots, g_n$. We define $\phi$ as follows.
For an element $g$ of $G$, we first write down the permutation *of the elements of $G$*

$$\begin{pmatrix} g_1 & \cdots & g_n \\ gg_1 & \cdots & gg_n \end{pmatrix}$$

obtained by left multiplying all the elements of $G$ by $g$.
Each product $gg_i$ must be $g_j$, for some $j$.
The element $\phi(g)$ of $S_n$ maps $i$ to $j$.
Formally, $\phi(g)$ is defined by

$$\phi(g) : i \mapsto j \quad \text{if and only if} \quad gg_i = g_j.$$

We must now show that $\phi$ satisfies three conditions:

(a) $\phi(g)$ is a uniquely defined element of $S_n$ for each $g$ in $G$, i.e. $\phi$ is a function from $G$ to $S_n$;

(b) $\phi$ is one–one;

(c) $\phi$ has the morphism property;

Firstly, by the way we have defined $\phi(g)$, it is uniquely determined by $g$ and maps the set
$$\{1, \ldots, n\}$$
to itself.

Furthermore, $\phi(g)$ is one–one. To see this, suppose that
$$\phi(g) : i \mapsto j \quad \text{and} \quad \phi(g) : k \mapsto j.$$
This requires that
$$gg_i = g_j \quad \text{and} \quad gg_k = g_j,$$
so that $gg_i = gg_k$. Left cancellation now gives $g_i = g_k$ and so $i = k$. Thus $\phi(g)$ is a one–one map from the finite set $\{1, \ldots, n\}$ to itself. It is, therefore, onto and so is a permutation of this set, that is
$$\phi(g) \in S_n.$$
This completes the proof that $\phi$ is a function from $G$ to $S_n$.

Secondly, $\phi$ is one–one. For, suppose that
$$\phi(g) = \phi(h),$$
for elements $g$ and $h$ of $G$. Since $\phi(g)$ belongs to $S_n$, we may assume that
$$\phi(g) : 1 \mapsto j,$$
for some $j$. But $\phi(g) = \phi(h)$ and so
$$\phi(h) : 1 \mapsto j.$$
By the definition of $\phi$, this means that
$$gg_1 = g_j \quad \text{and} \quad hg_1 = g_j.$$
Hence, by right cancellation, $g = h$, completing the proof that $\phi$ is one–one.

Finally, we tackle the morphism property. We need to show that, for any $g$ and $h$ in $G$,
$$\phi(gh) = \phi(g)\phi(h).$$
Since both $\phi(gh)$ and $\phi(g)\phi(h)$ are permutations in $S_n$, we show their equality by showing that they have the same effect on each of the elements in
$$\{1, \ldots, n\}.$$
We consider the effect of $\phi(gh)$ and $\phi(g)\phi(h)$ on
$$i \in \{1, \ldots, n\}.$$
Assume that
$$\phi(g) : j \mapsto k,$$
$$\phi(h) : i \mapsto j,$$
and hence that
$$\phi(g)\phi(h) : i \mapsto k.$$

Now, by the definition of $\phi$,

$$\phi(g): j \mapsto k \quad \text{if and only if} \quad gg_j = g_k$$

and

$$\phi(h): i \mapsto j \quad \text{if and only if} \quad hg_i = g_j.$$

By associativity in $G$,

$$\begin{aligned}(gh)g_i &= g(hg_i) \\ &= gg_j \\ &= g_k.\end{aligned}$$

But

$$(gh)g_i = g_k \quad \text{if and only if} \quad \phi(gh): i \mapsto k.$$

So

$$\phi(gh): i \mapsto k \quad \text{if and only if} \quad \phi(g)\phi(h): i \mapsto k,$$

for each $i$. Hence

$$\phi(gh) \quad \text{and} \quad \phi(g)\phi(h)$$

have the same effect on every element of $\{1, \ldots, n\}$. Thus

$$\phi(gh) = \phi(g)\phi(h)$$

and we have completed the proof of the morphism property.

As $\phi$ is a homomorphism, $\text{Im}(\phi)$ is a subgroup of $S_n$ and, as $\phi$ is one–one, it is an isomorphism from $G$ to $\text{Im}(\phi)$. ■

Had we remembered the formulation of the definition of a group action of a group $G$ on a set of $X$ as a homomorphism from $G$ to the group $\Gamma(X)$ of permutations of the set $X$, we could have reduced the last part of the proof to merely verifying that left multiplication does define an action of a group $G$ on itself.

Cayley's Theorem suggests that, to find out all about finite groups, one only needs to investigate the subgroups of the permutation groups $S_n$, for every positive integer $n$.

To see why this is less useful than might appear at first glance, consider the problem of finding all groups of order 10. Cayley's Theorem shows that all such groups appear as subgroups of $S_{10}$. Unfortunately, the order of $S_{10}$ is

$$10! = 3628800,$$

which suggests that it is probably easier to look for all groups of order 10 than to look for all subgroups of order 10 of this particular group of order 3628800.

In spite of these remarks, the *idea* of using homomorphisms from a group to a permutation group can produce useful results. We shall use a related technique in *Unit GR6*.

# 4 CONJUGACY

In Section 3 we showed that the group $A_4$, which has order 12, has no subgroup of order 6. This result shows that the converse of Lagrange's Theorem is false for non-Abelian groups. (As we have seen from Theorem 2.2, the converse is true for Abelian groups.)

After this negative result, we begin the investigation of which properties of Abelian groups *can* be generalized to non-Abelian groups.

In Section 5 of this unit we shall show that, for non-Abelian groups of prime power order, subgroups exist corresponding to every divisor of the order of the group. In *Unit GR5* we shall show that, for a restricted class of divisors of the order of a group, namely prime power divisors, corresponding subgroups always exist.

We shall restrict our discussion throughout to *finite* groups, although some of the methods generalize to infinite groups. Since we are no longer dealing exclusively with Abelian groups, we shall continue with the multiplicative notation that we reverted to in Section 3.

In Section 3 we discussed conjugacy in the permutation group $S_n$. In this section we investigate conjugacy in general and its link with the existence of normal subgroups. The aim is to develop some tools which we shall need for subsequent work.

We have already seen some examples of conjugacy being used to define an action of a group on its own underlying set. We now generalize this process. We shall make use of the Orbit–stabilizer Theorem from *Unit GE1*.

*Unit IB2.*

---

*Orbit–stabilizer theorem*

When a finite group $G$ acts on a finite set $X$, then for each $x \in X$:

$$|\operatorname{Orb}(x)| \times |\operatorname{Stab}(x)| = |G|.$$

---

As a consequence of this theorem we note that, for all $x \in X$,

$|\operatorname{Orb}(x)|$ divides $|G|$.

Since $|G| = |\operatorname{Orb}(x)| \times |\operatorname{Stab}(x)|$, we also know that $|\operatorname{Stab}(x)|$ divides $|G|$. As $\operatorname{Stab}(x)$ is a subgroup of $G$, we also knew this from Lagrange's Theorem.

We shall also use the fact that the orbits partition the set $X$. The reason for this fact is that, if we define a relation on the set $X$ by

$x$ is related to $y$ $\iff$ there is a $g \in G$ with $x = g \wedge y$,

then the definition of a group action ensures that this is an equivalence relation, and the equivalence classes (which partition $X$) are the orbits of the group action. Thus if

$\operatorname{Orb}(x_1), \ldots, \operatorname{Orb}(x_s)$

Here $s$ is the number of orbits.

are the distinct orbits of $X$ under the action of $G$, then

$|X| = |\operatorname{Orb}(x_1)| + \cdots + |\operatorname{Orb}(x_s)|.$

This equation, derived from the partition into orbits, will be referred to as the **partition equation** for the group action.

Because of the form of these results, our deductions will generally involve counting arguments and the divisibility properties of the integers.

As mentioned earlier, we consider the group action of $G$ on itself defined by conjugacy. This is a generalization of some of the examples that you met in *Unit IB2*.

*Exercise 4.1*

Let $G$ be a group. Show that

$$g \wedge x = gxg^{-1}$$

defines a group action of $G$ on (the underlying set of) $G$.

---

The element $gxg^{-1}$ is called the **conjugate** of $x$ by $g$, and $\operatorname{Orb}(x)$ for this action is called the **conjugacy class** of $x$, and consists of all conjugates of $x$.

Because of the importance of the result of Exercise 4.1, we make the following formal definition.

> **Definition 4.1 Conjugacy action**
>
> Let $G$ be a group. The action of $G$ on itself, i.e. on the underlying set of $G$, defined by
>
> $$g \wedge x = gxg^{-1}$$
>
> is called the **conjugacy action** (of $G$ on itself).

Since this conjugacy action is the only one that we shall discuss in this section, $\operatorname{Stab}(x)$ and $\operatorname{Orb}(x)$ will refer to this action throughout the remainder of this section.

The concept of conjugacy is useful in going from Abelian to non-Abelian groups because, in some sense, it gives a measure of how non-Abelian a group is. This remark is justified by observing that

$$gxg^{-1} = x \iff gx = xg.$$

In other words $x$ is equal to its conjugate by $g$ if and only if $x$ and $g$ commute.

This gives us an interpretation of $\operatorname{Stab}(x)$ for the conjugacy action:

$$\operatorname{Stab}(x) = \{g : g \text{ commutes with } x\}.$$

Thus the bigger $\operatorname{Stab}(x)$, the more of $G$ commutes with $x$.

The fact that the orbits partition the set being acted on, in this case $G$ itself, gives the following, very useful, result.

> **Theorem 4.1 Class equation**
>
> Let $G$ be a finite group and let
>
> $$\operatorname{Orb}(x_1), \ldots, \operatorname{Orb}(x_s)$$
>
> be the distinct orbits under the conjugacy action. Then
>
> $$|G| = |\operatorname{Orb}(x_1)| + \cdots + |\operatorname{Orb}(x_s)|,$$
>
> where
>
> $$|\operatorname{Orb}(x_i)| \text{ divides } |G|, \quad i = 1, \ldots, s.$$
>
> The equation
>
> $$|G| = |\operatorname{Orb}(x_1)| + \cdots + |\operatorname{Orb}(x_s)|,$$
>
> is called the **class equation** of $G$.

The class equation is a special case of the partition equation.

Although this theorem is a straightforward application of the Orbit–stabilizer Theorem and the partition equation, it is so useful that we paraphrase the result for emphasis.

> The conjugacy action splits a group into distinct conjugacy classes, each of whose orders divides the order of the group and the sum of these orders is the order of the group itself.

The following exercise is designed to give you some experience in writing down the class equation of groups.

*Exercise 4.2*

For each of the following groups, find its class equation:

(a) $C_4$;

(b) $V$, the Klein group;

(c) $S_3$;

(d) $S_4$.

We now look at the significance of the single-element conjugacy classes.

*Exercise 4.3*

Show that, for the conjugacy action,

$|\operatorname{Orb}(e)| = 1,$

where $e$ is the identity of the group $G$.

The solution of Exercise 4.3 and the Orbit–stabilizer Theorem show that

$\operatorname{Stab}(e) = G.$

This is equivalent to saying that every element of $G$ commutes with $e$.

More generally, $|\operatorname{Orb}(x)| = 1$ if and only if $\operatorname{Stab}(x) = G$, i.e. if and only if $x$ commutes with every element of $G$.

We know that there is always at least one single-element conjugacy class, namely the one containing $e$.

*Exercise 4.4*

Use the class equation to show that any group of order $9 = 3^2$ contains at least three single-element conjugacy classes.

This is a special case of a result that we shall prove in the next section.

Since single-element conjugacy classes correspond to elements which commute with all elements of the group, in an Abelian group every conjugacy class contains only one element and the class equation becomes

$$|G| = \overbrace{1 + \cdots + 1}^{|G| \text{ terms}}.$$

Abelian groups thus give one extreme case of the class equation and, conversely, if a group has such a class equation then it is Abelian.

We know that there is at least one 1 in the class equation. In some sense, the proportion of 1s in the class equation indicates 'how Abelian' the group is.

The 1 corresponds to the conjugacy class containing the identity.

The above discussion gives some indication of why we make the following definition.

> **Definition 4.2 Centre of a group**
>
> Let $G$ be a group. The **centre** of $G$ is the set $Z(G)$ defined by
>
> $Z(G) = \{x \in G : gx = xg, \text{ for all } g \in G\}$.
>
> In other words $Z(G)$ is the set of elements of $G$ which commute with every element in $G$.
> Equivalently $Z(G)$ is the union of the single-element conjugacy classes.

From the definition of $Z(G)$, the group $G$ is Abelian if and only if $G = Z(G)$.

Exercise 4.4 shows that the centre of a group of order 9 has at least 3 elements.

*Exercise 4.5*

Prove that, for any group $G$, $Z(G)$ is a normal subgroup of $G$.

You must first establish that $Z(G)$ is a *subgroup* of $G$.

*Exercise 4.6*

Prove that any subgroup $H$ of $Z(G)$ is a normal subgroup of the group $G$.

Because $Z(G)$ is a subgroup of $G$ and is the union of the single-element conjugacy classes, Lagrange's Theorem tells us that the number of single-element conjugacy classes is a divisor of the order of the group. This last result is useful enough to be worth noting as a theorem.

> **Theorem 4.2**
>
> Let $G$ be a finite group. Then the number of 1s in the class equation of $G$ is the order of $Z(G)$, the centre of $G$, and is a divisor of $|G|$.

We now know that, for any group $G$, the centre $Z(G)$ is a normal subgroup of $G$. It is reasonable to ask whether information regarding the quotient group $G/Z(G)$ provides information about $G$ itself. We ask you to prove one such result in the following exercise.

*Exercise 4.7*

Show that if the quotient group

$G/Z(G)$,

of a group $G$ by its centre, is cyclic, then $G$ is Abelian.

*Hint* Take the quotient group $G/Z(G)$ to be generated by the coset $aZ(G)$. Consider the form of all elements of $G/Z(G)$, i.e. the form of all cosets of $Z(G)$, the fact that the cosets of $Z(G)$ partition $G$ and the definition of $Z(G)$.

*Exercise 4.8*

Deduce from the results of Exercises 4.4 and 4.7 that there are only two groups of order 9.

*Hint* Make use of the Canonical Decomposition Theorem for Finitely Generated Abelian Groups.

*Exercise 4.9*

Prove that a group of order 15 is either Abelian or has a trivial centre.

In fact, as we shall show in *Unit GR5*, all groups of order 15 are Abelian.

The result of Exercise 4.7 is of sufficient importance to be stated as the following theorem.

> **Theorem 4.3**
>
> Let $G$ be a group and let $Z(G)$ be the centre of $G$.
> If the quotient group $G/Z(G)$ is cyclic, then the group $G$ is Abelian.

This theorem means that the only way in which the quotient group $G/Z(G)$ can be cyclic is for $Z(G)$ to be the whole of $G$. Thus, if $G/Z(G)$ is cyclic then it is the trivial cyclic group.

The main use we shall make of Theorem 4.3 is to rule out various possibilities when building a catalogue of non-Abelian groups of a particular order.

The union of the one-element conjugacy classes of a group $G$ produces a particular normal subgroup of $G$, the centre. There is a more general link between conjugacy classes of $G$ and normal subgroups, which is given by the following theorem.

> **Theorem 4.4**
>
> Let $H$ be a subgroup of a group $G$.
> Then $H$ is normal if and only if $H$ is a union of conjugacy classes of $G$.

### Proof

*If*

Suppose that $H$ is a union of conjugacy classes of $G$.
Therefore, for any $h \in H$, we have $h \in \mathrm{Orb}(h) \subseteq H$.
Now, for every $g \in G$ we have

$$g \wedge h = ghg^{-1} \in \mathrm{Orb}(h) \subseteq H.$$

Therefore

$$gHg^{-1} \subseteq H$$

and $H$ is a normal subgroup of $G$.

*Only if*

Suppose $H$ is normal in $G$.
For every $g \in G$ we have

$$gHg^{-1} \subseteq H.$$

Hence, for each $x \in H$, every conjugate $gxg^{-1}$ of $x$ is also in $H$.
Thus, if $x \in H$, then $\mathrm{Orb}(x) \subseteq H$.
Therefore $H$ is a union of orbits, i.e. of conjugacy classes, of $G$. ∎

Theorem 4.4 shows that the class equation for a group $G$ places restrictions on the possible orders of normal subgroups of $G$ in addition to those imposed by Lagrange's Theorem. The order of a normal subgroup must be the sum of the orders of some conjugacy classes. Furthermore, one of these conjugacy classes must be $\{e\}$, which has order 1.

**Example 4.1**

As you found earlier, the class equation for $S_3$ is

$6 = 1 + 3 + 2$.

By Theorem 4.4, the possible orders for normal subgroups are:

$1 = 1;$
$1 + 3 = 4;$
$1 + 2 = 3;$
$1 + 3 + 2 = 6.$

Of these, 1 corresponds to the trivial normal subgroup. Lagrange's Theorem rules out 4. The whole group corresponds to 6.

The subgroup $\{e, (123), (132)\}$, corresponding to 3, has index 2 in $S_3$ and so is normal.

*Unit IB4, Exercise 3.4.*

Hence, *in this case*, there is a normal subgroup of each order satisfying both the class equation condition and Lagrange's Theorem.

On the other hand, we know that $S_3$ has subgroups of order 2, for example $\{e, (12)\}$. So the class equation provides a proof that such subgroups cannot be normal. ♦

In practice, we usually use Theorem 4.4 to prove the non-existence of normal subgroups of a particular order. We do so by showing that no collection of conjugacy classes, *including* $\{e\}$, can give the required number of elements.

*All subgroups must include $e$.*

On the other hand, even if a sum of terms in the class equation satisfies Lagrange's Theorem and any other necessary conditions relating to the possible orders of subgroups, the union of the corresponding conjugacy classes might not form a subgroup. However, if the union is a subgroup then, by Theorem 4.4, it will be normal.

*Exercise 4.10*

Show that a group $G$ of order 60, with class equation

$60 = 1 + 12 + 12 + 15 + 20$,

has no non-trivial proper normal subgroups.

*A proper subgroup is one which is not equal to the whole group.*

As a final, practice exercise using the ideas of this section, we ask you to investigate conjugacy in the dihedral group $D_5$.

*Exercise 4.11*

Let $D_5$ be defined by the presentation

$D_5 = \langle a, b : a^5 = e, \ b^2 = e, \ ba = a^{-1}b \, (= a^4 b) \rangle.$

Thus,

$D_5 = \{e, a, a^2, a^3, a^4, b, ab, a^2 b, a^3 b, a^4 b\}.$

(a) Show that

$ba^j = a^{-j} b, \quad j = 2, 3, 4.$

(b) Find the conjugacy classes of $D_5$ and, hence, its class equation.

(c) Show that $D_5$ has trivial centre, no normal subgroup of order 2 and a normal subgroup of order 5.

*We have chosen to deal with $D_5$ entirely algebraically, using the presentation given. There is another approach which uses a geometric interpretation of conjugacy and of dihedral groups. At this point, discussing such an approach would be a diversion from the main purpose of this section.*

# 5 $p$-GROUPS

In this section we apply some of the results from Section 4 to a particular class of groups: those whose order is a power of a prime. We shall use the class equation and the fact that, if the quotient group

$$G/Z(G)$$

of $G$ by its centre is cyclic, then $G$ is Abelian.

In the course of our investigations, we shall need to prove a theorem about quotient groups which has general use.

This section continues the theme of seeing what can be said about groups if various restrictions are placed upon them.

---

**Definition 5.1 $p$-Group**

Let $p$ be a prime and $n$ a positive integer.
Then a group $G$ of order $p^n$ is called a $p$-group.

---

In this section, $p$ will always denote a prime number.

In Section 4 we considered one special case: we looked at groups of order $9 = 3^2$. We found that the centre of such a group had to be non-trivial since it had to have order at least 3. We were then able to deduce that a group of order 9 must be Abelian. It follows that the only groups of order 9 are $\mathbb{Z}_9$ and $\mathbb{Z}_3 \times \mathbb{Z}_3$.

We now generalize these observations. Firstly, we ask you to show that the centre of *any* $p$-group is non-trivial. Secondly, we ask you to deduce that any group of order $p^2$ is Abelian.

*Exercise 5.1*

Let $G$ be a $p$-group of order $p^n$. Use the class equation to show that $Z(G)$, the centre of $G$, has order at least $p$.

*Exercise 5.2*

Let $G$ be a $p$-group of order $p^2$. Show that $G$ is Abelian.

---

The solution to Exercise 5.2 means that we know everything about groups of order $p^2$, for any prime $p$. Since they are all Abelian, the only possibilities, by the Canonical Decomposition Theorem, are

$$\mathbb{Z}_{p^2} \quad \text{and} \quad \mathbb{Z}_p \times \mathbb{Z}_p.$$

We now try to generalize our observations about groups of order $p^2$. There are two directions we can take. The first, which we consider in this unit, is to increase the exponent. The second, which we discuss in *Unit GR5*, is to consider the product of two *different* primes.

Whereas groups of order $p^2$ must be Abelian, the situation is rather more complicated for groups of order $p^3$. For the case $p = 2$, our survey in *Unit GR2* produced three Abelian and two non-Abelian groups of order $2^3 = 8$. The non-Abelian ones are the dihedral group $D_4$ and Hamilton's quaternion group $Q$.

The fact that the centre of a $p$-group must be non-trivial does give *some* information about groups of order $p^3$. The centre $Z(G)$ of such a group $G$ may only be of order $p$, $p^2$ or $p^3$.

*Exercise 5.3*

Let $G$ be a group of order $p^3$. Show that $Z(G)$ cannot be of order $p^2$.

---

We have now reduced the possibilities for $|Z(G)|$ to $p$ and $p^3$.

If the order of $Z(G)$ is $p^3$, the group is Abelian and we know, by the Canonical Decomposition Theorem, that the only possibilities are

$$\mathbb{Z}_{p^3}, \quad \mathbb{Z}_p \times \mathbb{Z}_{p^2} \quad \text{and} \quad \mathbb{Z}_p \times \mathbb{Z}_p \times \mathbb{Z}_p.$$

If the order of $Z(G)$ is $p$, then the quotient group $G/Z(G)$ has order $p^2$ and is, by Exercise 5.2, Abelian, and so may only be one of

$$\mathbb{Z}_{p^2} \quad \text{or} \quad \mathbb{Z}_p \times \mathbb{Z}_p.$$

However, if the quotient were $\mathbb{Z}_{p^2}$, we would have a cyclic quotient, $G$ would be Abelian and the order of the centre would be $p^3$ and not $p$.
We conclude that a non-Abelian group $G$ of order $p^3$ has a centre of order $p$ and that

$$G/Z(G) \cong \mathbb{Z}_p \times \mathbb{Z}_p.$$

Unfortunately, this information does not uniquely determine the structure of $G$. For example, this discussion shows that any non-Abelian group $G$ of order 8 has a centre of order 2 and that the quotient by the centre is $\mathbb{Z}_2 \times \mathbb{Z}_2$. However, we know that there are two different non-Abelian groups of order 8, and they must both satisfy these conditions.

This example shows that knowing the structure of a normal subgroup and the structure of the corresponding quotient group is not, in general, sufficient to determine the structure of the group itself.

Despite the fact that $p$-groups may be non-Abelian, they share a subgroup property with Abelian groups: subgroups exist corresponding to each divisor of the order of the group. In fact, a little more is true and the full result is contained in the following theorem.

---

**Theorem 5.1**

Let $G$ be a $p$-group of order $p^n$ for some positive integer $n$.
Then there exist normal subgroups

$$\{e\} = H_0, H_1, \ldots, H_n = G$$

of $G$ such that

$$\{e\} = H_0 \subset H_1 \subset \cdots \subset H_{n-1} \subset H_n = G,$$

where $|H_i| = p^i$, $i = 0, \ldots, n$.

---

Because of the set inclusions involving the $H_i$s, we say that they form a *chain* of subgroups.

*Proof*

The proof uses the Principle of Mathematical Induction for $n$.

For $n = 1$, $|G| = p$ and the required chain is simply

$$\{e\} = H_0 \subset H_1 = G.$$

We have $|H_i| = p^i$ for $i = 0, 1$ and the normality condition is trivially true.

Now assume that the theorem is true for all groups of order $p^r$, where $r \leq k$, and let $G$ have order $p^{k+1}$.

If $G$ is Abelian, then, since $p^k$ divides $p^{k+1}$, $G$ has a subgroup $H_k$ of order $p^k$. By the induction hypothesis, $H_k$ has a chain of subgroups with the correct orders:

$$\{e\} = H_0 \subset H_1 \subset \cdots \subset H_{k-1} \subset H_k.$$

Combining this with $H_k \subset H_{k+1} = G$, we have the required chain:
$$\{e\} = H_0 \subset H_1 \subset \cdots \subset H_k \subset H_{k+1} = G.$$
Since $G$ is Abelian, normality is automatic.

Now we consider the case where $G$ is definitely *not* Abelian. Our strategy is to use what we know about the centre of a $p$-group to construct the chain of subgroups in two stages: from the trivial subgroup to the centre and then from the centre to the whole group.

Since $G$ is non-Abelian, its centre $Z(G)$ is not the whole group and so its order is *less* than $p^{k+1}$. By Lagrange's Theorem, $Z(G)$ is a $p$-group. So,
$$|Z(G)| = p^s, \quad 0 \le s \le k,$$
and we may apply the induction hypothesis to $Z(G)$. Thus, there exists a chain of subgroups
$$\{e\} = H_0 \subset H_1 \subset \cdots \subset H_s = Z(G)$$
of $Z(G)$ where $|H_i| = p^i$, $i = 0, \ldots, s$. These subgroups are also subgroups of $G$. Furthermore, by Exercise 4.6, they are normal subgroups of $G$.

We now have the part of the required chain from the trivial subgroup to $Z(G)$, and we turn to completing the chain from $Z(G)$ to $G$.

To apply the induction hypothesis, we need another $p$-group of order less than $p^{k+1}$. The one we shall use is the quotient of $G$ by its centre. Since $G$ is a $p$-group, by Exercise 5.1 we know that $Z(G)$ has at least $p$ elements. Hence
$$|Z(G)| = p^s, \quad 1 \le s \le k.$$
Since $Z(G)$ is normal in $G$, we may consider the quotient group $G/Z(G)$, whose order is
$$p^{k+1}/p^s = p^{k+1-s} \le p^k.$$
Hence, the quotient is a $p$-group to which we may apply the induction hypothesis.

It follows that there exists a chain of normal subgroups of the quotient,
$$\{Z(G)\} = \overline{H}_0 \subset \overline{H}_1 \subset \cdots \subset \overline{H}_{k+1-s} = G/Z(G),$$
where $|\overline{H}_i| = p^i$, $i = 0, \ldots, k+1-s$.

To complete the proof, we must show that the chain of normal subgroups in the quotient group $G/Z(G)$ may be used to define a chain of normal subgroups of $G$ between $Z(G)$ and $G$. The fact that we can do so is an immediate consequence of the following theorem. □

---

**Theorem 5.2 Correspondence theorem**

Let $G$ be a group and let $N$ be a normal subgroup of $G$.
Then there is a one–one correspondence between subgroups of $G$ which contain $N$ and subgroups of the quotient group $G/N$.
In particular, $N$ corresponds to the trivial subgroup of the quotient group, and $G$ to the whole of the quotient group.
Furthermore, in this correspondence, normal subgroups correspond to normal subgroups.

---

For the moment we take this theorem on trust and use it to complete our proof.

### Proof of Theorem 5.1 continued

By Theorem 5.2, each normal subgroup $\overline{H}_i$ of $G/Z(G)$ gives rise to a normal subgroup $H_{i+s}$ of $G$ which contains $Z(G)$. Since the correspondence is one–one and the $\overline{H}_i$s are all distinct, so are the $H_{i+s}$s. Also, the trivial subgroup $\overline{H}_0$ of the quotient group corresponds to $Z(G) = H_s$.

We shall see from the proof of Theorem 5.2 that set inclusion is preserved by the correspondence. Hence, we have a chain of normal subgroups

$$Z(G) = H_s \subset H_{s+1} \subset \cdots \subset H_{k+1} = G.$$

Furthermore, since $|Z(G)| = p^s$ and $|G| = p^{k+1}$ and since there are $k - s$ subgroups in the (strictly inclusive) chain *between* $H_s = Z(G)$ and $H_{k+1} = G$, each of order $p^i$ for some $i$, we must have $|H_i| = p^i$, for $i = s, \ldots, k+1$.

We have now constructed a chain of normal subgroups

$$\{e\} = H_0 \subset H_1 \subset \cdots \subset H_s = Z(G) = H_s \subset \cdots \subset H_{k+1} = G$$

for $G$, where $|H_i| = p^i$, $i = 0, \ldots, k+1$, completing the inductive step. ∎

We have now to provide a proof of Theorem 5.2.

The proof falls naturally into several stages.

(a) We define a function $\alpha$ from the set of subgroups of $G$ containing $N$ to the set of subgroups of $G/N$.

(b) We define a function $\beta$ from the set of subgroups of $G/N$ to the set of subgroups of $G$ containing $N$.

(c) We show that $\alpha$ is one–one and onto. The proof that it is onto uses the function $\beta$.

(d) We show that normal subgroups correspond to normal subgroups.

The other results in the theorem will follow from the details of the way that the two functions are defined, as will the fact that the correspondence preserves inclusion.

### Proof of Theorem 5.2

From the statement of the theorem, $G$ is a group and $N$ is a normal subgroup of $G$.

We ask you to provide the first two parts of the proof in the following exercise. □

### Exercise 5.4

(a) Let $H$ be a subgroup of $G$ containing $N$. That is,

$$N \subseteq H \subseteq G.$$

Let $\alpha(H)$ be the set

$$\alpha(H) = \overline{H} = \{hN : h \in H\}.$$

Prove that $\alpha(H)$ is a subgroup of $G/N$.

$\overline{H}$ is the image of $H$ under the natural homomorphism from $G$ to $G/N$.

(b) Let $\overline{H}$ be a subgroup of $G/N$. Let $\beta(\overline{H})$ be the set

$$\beta(\overline{H}) = H = \{h \in G : hN \in \overline{H}\}.$$

Prove that $\beta(\overline{H})$ is a subgroup of $G$ containing $N$.

### Proof of Theorem 5.2 continued

As a result of Exercise 5.4 we have:

(a) a function $\alpha$ from the set of subgroups of $G$ which contain $N$ to the set of subgroups of the quotient group $G/N$, where $\alpha$ preserves inclusion;

(b) a function $\beta$ from the set of subgroups of $G/N$ to the set of subgroups of $G$ which contain $N$.

We now prove that $\alpha$ is one–one.
Suppose that $H_1$ and $H_2$ are two subgroups of $G$ containing $N$, such that

$$\alpha(H_1) = \alpha(H_2).$$

Let $x_1$ be an element of $H_1$.
By the definition of $\alpha$, the coset $x_1 N$ is an element of $\alpha(H_1) = \alpha(H_2)$.
Hence, there exists an element $x_2$ of $H_2$ such that

$$x_1 N = x_2 N.$$

Therefore, by the condition for the equality of cosets,

$$x_2^{-1} x_1 \in N.$$

Now, $N$ is contained in $H_2$ and so

$$x_2^{-1} x_1 \in H_2.$$

But, $x_2 \in H_2$ and so, by closure,

$$x_2(x_2^{-1} x_1) = x_1 \in H_2.$$

This shows that

$$H_1 \subseteq H_2.$$

However, the conditions are symmetric in $H_1$ and $H_2$ and so, interchanging the subscripts 1 and 2 throughout, gives a proof that

$$H_2 \subseteq H_1.$$

Hence, $H_1 = H_2$ and $\alpha$ is one–one.

We now use $\beta$ to show that $\alpha$ is onto.
Let $\overline{H}$ be a subgroup of $G/N$. By Exercise 5.4, we know that $H = \beta(\overline{H})$ is a subgroup of $G$ containing $N$.
We now show that $\alpha(H) = \alpha(\beta(\overline{H})) = \overline{H}$ to verify that $\alpha$ is onto.
We have

$$\alpha(H) = \{hN : h \in H\}.$$

But, from the definition of $\beta$, we have $h \in H$ if and only if $hN \in \overline{H}$. Hence,

$$\alpha(H) = \overline{H}.$$

We have established that the correspondence, given by $\alpha$, exists, is one–one and onto, and preserves inclusion. We can also deduce from the definition of $\alpha$ that $\alpha(N) = N$, the trivial subgroup of $G/N$, and that $\alpha(G) = G/N$, the whole of the quotient group.

*The proof has also established that $\alpha$ and $\beta$ are inverses of each other.*

The remaining part of the theorem, namely that normal subgroups correspond to normal subgroups, is set as an exercise. □

### Exercise 5.5

(a) Let $H$ be a normal subgroup of $G$ which contains $N$.
    Show that $\alpha(H) = \overline{H}$ is a normal subgroup of $G/N$.
(b) Let $\overline{H}$ be a normal subgroup of $G/N$.
    Show that $\beta(\overline{H}) = H$ is a normal subgroup of $G$ which contains $N$.

### Proof of Theorem 5.2 continued

As a result of Exercise 5.5, we know that in the correspondence defined by $\alpha$ (and its inverse $\beta$), normal subgroups correspond to normal subgroups. This concludes the proof. ∎

Thus, by Theorem 5.1, a $p$-group $G$ has (normal) subgroups corresponding to every divisor of the order of $G$.

# SOLUTIONS TO THE EXERCISES

**Solution 1.1**

(a) We
$$900 = 2^2 \times 3^2 \times 5^2.$$
Arguing as in the example, we obtain the following table.

| prime power | factors of $d_{k-1}$ | $d_k$ | label |
|---|---|---|---|
| $5^2$ |  | $5^2$ | 5a |
|  | 5 | 5 | 5b |
| $3^2$ |  | $3^2$ | 3a |
|  | 3 | 3 | 3b |
| $2^2$ |  | $2^2$ | 2a |
|  | 2 | 2 | 2b |

Hence, there are $2 \times 2 \times 2 = 8$ Abelian groups of order 900.

| | |
|---|---|
| $\mathbb{Z}_{900}$ | 5a, 3a, 2a |
| $\mathbb{Z}_2 \times \mathbb{Z}_{450}$ | 5a, 3a, 2b |
| $\mathbb{Z}_3 \times \mathbb{Z}_{300}$ | 5a, 3b, 2a |
| $\mathbb{Z}_6 \times \mathbb{Z}_{150}$ | 5a, 3b, 2b |
| $\mathbb{Z}_5 \times \mathbb{Z}_{180}$ | 5b, 3a, 2a |
| $\mathbb{Z}_{10} \times \mathbb{Z}_{90}$ | 5b, 3a, 2b |
| $\mathbb{Z}_{15} \times \mathbb{Z}_{60}$ | 5b, 3b, 2a |
| $\mathbb{Z}_{30} \times \mathbb{Z}_{30}$ | 5b, 3b, 2b |

(b) We have
$$432 = 2^4 \times 3^3.$$
The corresponding table is as follows.

| prime power | factors of $d_{k-3}$ | $d_{k-2}$ | $d_{k-1}$ | $d_k$ | label |
|---|---|---|---|---|---|
| $3^3$ |  |  |  | $3^3$ | 3a |
|  |  |  | 3 | $3^2$ | 3b |
|  |  | 3 | 3 | 3 | 3c |
| $2^4$ |  |  |  | $2^4$ | 2a |
|  |  |  | 2 | $2^3$ | 2b |
|  |  |  | $2^2$ | $2^2$ | 2c |
|  |  | 2 | 2 | $2^2$ | 2d |
|  | 2 | 2 | 2 | 2 | 2e |

Hence, there are $3 \times 5 = 15$ Abelian groups of order 432.

| | |
|---|---|
| $\mathbb{Z}_{432}$ | 3a, 2a |
| $\mathbb{Z}_2 \times \mathbb{Z}_{216}$ | 3a, 2b |
| $\mathbb{Z}_4 \times \mathbb{Z}_{108}$ | 3a, 2c |
| $\mathbb{Z}_2 \times \mathbb{Z}_2 \times \mathbb{Z}_{108}$ | 3a, 2d |
| $\mathbb{Z}_2 \times \mathbb{Z}_2 \times \mathbb{Z}_2 \times \mathbb{Z}_{54}$ | 3a, 2e |
| $\mathbb{Z}_3 \times \mathbb{Z}_{144}$ | 3b, 2a |
| $\mathbb{Z}_6 \times \mathbb{Z}_{72}$ | 3b, 2b |
| $\mathbb{Z}_{12} \times \mathbb{Z}_{36}$ | 3b, 2c |
| $\mathbb{Z}_2 \times \mathbb{Z}_6 \times \mathbb{Z}_{36}$ | 3b, 2d |
| $\mathbb{Z}_2 \times \mathbb{Z}_2 \times \mathbb{Z}_6 \times \mathbb{Z}_{18}$ | 3b, 2e |
| $\mathbb{Z}_3 \times \mathbb{Z}_3 \times \mathbb{Z}_{48}$ | 3c, 2a |
| $\mathbb{Z}_3 \times \mathbb{Z}_6 \times \mathbb{Z}_{24}$ | 3c, 2b |
| $\mathbb{Z}_3 \times \mathbb{Z}_{12} \times \mathbb{Z}_{12}$ | 3c, 2c |
| $\mathbb{Z}_6 \times \mathbb{Z}_6 \times \mathbb{Z}_{12}$ | 3c, 2d |
| $\mathbb{Z}_2 \times \mathbb{Z}_6 \times \mathbb{Z}_6 \times \mathbb{Z}_6$ | 3c, 2e |

**Solution 1.2**

Drawing up a table, we obtain the following.

| prime power | factors of | | | | |
|---|---|---|---|---|---|
| | $d_{k-4}$ | $d_{k-3}$ | $d_{k-2}$ | $d_{k-1}$ | $d_k$ |
| $p^5$ | | | | | $p^5$ |
| | | | | $p$ | $p^4$ |
| | | | | $p^2$ | $p^3$ |
| | | | $p$ | $p$ | $p^3$ |
| | | | $p$ | $p^2$ | $p^2$ |
| | | $p$ | $p$ | $p$ | $p^2$ |
| | $p$ | $p$ | $p$ | $p$ | $p$ |

Hence there are 7 Abelian groups of order $p^5$.

**Solution 1.3**

(a) Comparing $p^3 q^2 r$ with
$$360 = 2^3 \times 3^2 \times 5,$$
we see that exactly the same arguments as in Example 1.1 tell us that there are 6 different Abelian groups of order $p^3 q^2 r$.

*All that we use the inequality $p < q < r$ for is to ensure that the three primes are distinct.*

(b) Comparing $p^2 q^2 r^2$ with $900 = 2^2 \times 3^2 \times 5^2$ shows that there are 8 Abelian groups of order $p^2 q^2 r^2$.

(c) Comparing $p^4 q^3$ with $432 = 2^4 \times 3^3$ shows that there are 15 Abelian groups of order $p^4 q^3$.

**Solution 2.1**

All three non-identity elements of the group, namely

$$(1,0), \quad (0,1) \quad \text{and} \quad (1,1),$$

have order 2.

**Solution 2.2**

(a) Since $\mathbb{Z}_3 \times \mathbb{Z}_3$ has order 9, by Lagrange's Theorem the only possible orders of elements are 1, 3 and 9. But the group is not cyclic, since 3 and 3 are not coprime, so the possibilities are only 1 and 3. As any group has only one element of order 1, namely the identity, all eight non-identity elements must have order 3.

(b) Each non-identity element has order 3 and generates a cyclic subgroup of order 3 containing the identity and two elements of order 3. Furthermore, any two such subgroups of order 3 are either identical or intersect in just the identity. (This follows from Lagrange's Theorem, because the intersection of two subgroups must be a subgroup and so have order 1 or 3.)
Each subgroup of order 3 is uniquely defined by an element of order 3 and its inverse.
Since there are eight non-identity elements, all of order 3, there must be four subgroups of order 3.

**Solution 2.3**

We work as in Example 2.1. First, we write down the prime decomposition of the torsion coefficients.

$$2 = 2$$
$$6 = 2 \times 3$$
$$30 = 2 \times 3 \times 5$$
$$150 = 2 \times 3 \times 5^2$$

Next, we decompose the terms in the canonical direct product.

$$\mathbb{Z}_2 = \mathbb{Z}_2$$
$$\mathbb{Z}_6 \cong \mathbb{Z}_2 \times \mathbb{Z}_3$$
$$\mathbb{Z}_{30} \cong \mathbb{Z}_2 \times \mathbb{Z}_3 \times \mathbb{Z}_5$$
$$\mathbb{Z}_{150} \cong \mathbb{Z}_2 \times \mathbb{Z}_3 \times \mathbb{Z}_{5^2}$$

Finally, we assemble the information and reorder the terms.

$$A = \mathbb{Z}_2 \times \mathbb{Z}_6 \times \mathbb{Z}_{30} \times \mathbb{Z}_{150}$$
$$\cong \mathbb{Z}_2 \times (\mathbb{Z}_2 \times \mathbb{Z}_3) \times (\mathbb{Z}_2 \times \mathbb{Z}_3 \times \mathbb{Z}_5) \times (\mathbb{Z}_2 \times \mathbb{Z}_3 \times \mathbb{Z}_{5^2})$$
$$\cong (\mathbb{Z}_2 \times \mathbb{Z}_2 \times \mathbb{Z}_2 \times \mathbb{Z}_2) \times (\mathbb{Z}_3 \times \mathbb{Z}_3 \times \mathbb{Z}_3) \times (\mathbb{Z}_5 \times \mathbb{Z}_{5^2})$$

Hence, the 2-primary component is

$$\mathbb{Z}_2 \times \mathbb{Z}_2 \times \mathbb{Z}_2 \times \mathbb{Z}_2,$$

the 3-primary component is

$$\mathbb{Z}_3 \times \mathbb{Z}_3 \times \mathbb{Z}_3$$

and the 5-primary component is

$$\mathbb{Z}_5 \times \mathbb{Z}_{5^2}.$$

**Solution 2.4**

The group $A$ has only one primary component, the 2-primary component $A$ itself. The order of this primary component is

$$2^5 = 2 \times 2 \times 8 = 2^1 \times 2^1 \times 2^3,$$

corresponding to the cyclic components in the decomposition.

(a) We are looking for subgroups of order $4 = 2^2$. The exponent 2, in the order of the subgroup, may be written in a number of ways as a sum of three terms, each less than or equal to the corresponding exponent in the 2-primary component:

$$2 = 0 + 0 + 2$$
$$= 0 + 1 + 1$$
$$= 1 + 0 + 1$$
$$= 1 + 1 + 0.$$

Corresponding to $0 + 0 + 2$ we select the trivial subgroup of each of the first two components and the (unique cyclic) subgroup $\{0, 2, 4, 6\} \cong \mathbb{Z}_4$ of order $2^2 = 4$ from $\mathbb{Z}_8$. This gives rise to a subgroup

$$B = \{0\} \times \{0\} \times \mathbb{Z}_4 \cong \mathbb{Z}_4.$$

Corresponding to $0 + 1 + 1$ we select the trivial subgroup from the first component, the subgroup of $\mathbb{Z}_2$ of order $2^1 = 2$, i.e. $\mathbb{Z}_2$ itself, from the second and the (unique cyclic) subgroup $\{0, 4\} \cong \mathbb{Z}_2$ of order $2^1 = 2$ from $\mathbb{Z}_8$. This gives rise to a subgroup

$$C = \{0\} \times \mathbb{Z}_2 \times \mathbb{Z}_2 \cong \mathbb{Z}_2 \times \mathbb{Z}_2,$$

the Klein group.

Since $B \not\cong C$, we have the required two non-isomorphic subgroups of order 4.

> We know from *Unit GR2* that there are only two different groups of order 4, both of which are Abelian. This solution shows that both arise as subgroups of $A$.

(If we had chosen $1 + 0 + 1$ or $1 + 1 + 0$, we would have obtained

$$\mathbb{Z}_2 \times \{0\} \times \mathbb{Z}_2 \quad \text{or} \quad \mathbb{Z}_2 \times \mathbb{Z}_2 \times \{0\},$$

respectively, each of which is also the Klein group.)

(b) We are looking for subgroups of order $8 = 2^3$. The exponent 3, in the order of the subgroup, may be written in a number of ways as a sum of three terms, each less than or equal to the corresponding exponent in the 2-primary component:

$$3 = 0 + 0 + 3$$
$$= 0 + 1 + 2$$
$$= 1 + 0 + 2$$
$$= 1 + 1 + 1.$$

Corresponding to $0 + 0 + 3$ we select the trivial subgroup of the first two components and the group $\mathbb{Z}_8$ itself from the third. This gives rise to a subgroup

$$B = \{0\} \times \{0\} \times \mathbb{Z}_8 \cong \mathbb{Z}_8.$$

Corresponding to $0 + 1 + 2$ we select the trivial subgroup of the first component, the group $\mathbb{Z}_2$ itself from the second and the (unique cyclic) subgroup $\{0, 2, 4, 6\} \cong \mathbb{Z}_4$ of order $2^2 = 4$ from $\mathbb{Z}_8$. This gives rise to the subgroup

$$C = \{0\} \times \mathbb{Z}_2 \times \mathbb{Z}_4 \cong \mathbb{Z}_2 \times \mathbb{Z}_4.$$

The choice corresponding to $1 + 0 + 2$ again gives rise to the subgroup

$$\mathbb{Z}_2 \times \mathbb{Z}_4,$$

so this gets us no further.

Lastly, corresponding to $1 + 1 + 1$, we select the groups $\mathbb{Z}_2$ themselves from the first two components and the (unique cyclic) subgroup $\{0, 4\} \cong \mathbb{Z}_2$ from the third. The corresponding subgroup is

$$D \cong \mathbb{Z}_2 \times \mathbb{Z}_2 \times \mathbb{Z}_2.$$

Since $B$, $C$ and $D$ are non-isomorphic, we have found the required three non-isomorphic subgroups of order 8.

> We know from *Unit GR2* that there are only three different Abelian groups of order 8. This solution shows that all three arise as subgroups of $A$.

**Solution 2.5**

Using the notation of the proof of Theorem 2.2, the group $A$ has a 3-primary component

$$A_1 = \mathbb{Z}_3 \times \mathbb{Z}_9$$

and a 5-primary component

$$A_2 = \mathbb{Z}_{25}.$$

We require a subgroup $B$ of order $45 = 3^2 \times 5^1$. Using the strategy of the theorem, we find a subgroup $B_1 \subseteq A_1$ of order $3^2$ and a subgroup $B_2 \subseteq A_2$ of order 5.

Dealing with $B_1$ first, the possibilities for $B_1$ are

$$\{0\} \times \mathbb{Z}_9 \cong \mathbb{Z}_9$$

and

$$\mathbb{Z}_3 \times \{0, 3, 6\} \cong \mathbb{Z}_3 \times \mathbb{Z}_3.$$

The only possibility for $B_2$ is

$$\{0, 5, 10, 15, 20\} \cong \mathbb{Z}_5.$$

Hence, there are two subgroups of order 45. One is isomorphic to

$$\mathbb{Z}_9 \times \mathbb{Z}_5 \cong \mathbb{Z}_{45},$$

which is cyclic; the other is isomorphic to

$$\mathbb{Z}_3 \times \mathbb{Z}_3 \times \mathbb{Z}_5 \cong \mathbb{Z}_3 \times \mathbb{Z}_{15},$$

which is not cyclic.

### Solution 3.6

First, we write each permutation in cycle form:

$$x = (1346)(25);$$
$$y = (16)(2345)$$
$$= (2345)(16).$$

Since $x$ and $y$ have the same cycle type, they are conjugate.

A suitable $g$ obtained from the above cycle decompositions is

$$g = \begin{pmatrix} 1 & 3 & 4 & 6 & 2 & 5 \\ 2 & 3 & 4 & 5 & 1 & 6 \end{pmatrix}$$
$$= (12)(65).$$

Frame 11.

### Solution 3.7

Using the renaming given in the question, the Cayley table for $G$ is as follows:

|       | $g_1$ | $g_2$ | $g_3$ |
|-------|-------|-------|-------|
| $g_1$ | $g_1$ | $g_2$ | $g_3$ |
| $g_2$ | $g_2$ | $g_3$ | $g_1$ |
| $g_3$ | $g_3$ | $g_1$ | $g_2$ |

Hence,

$$\phi(g_1) = e,$$
$$\phi(g_2) = (123),$$
$$\phi(g_3) = (132).$$

The image of $\phi$ is the subset

$$\{e, (123), (132)\}$$

of $S_3$. Since

$$(123)^2 = (132) \quad \text{and} \quad (123)^3 = e,$$

this subset is a cyclic group of order 3. It is, therefore, isomorphic to $G$.

### Solution 4.1

We check the three requirements in turn.

Firstly, for all $g \in G$, $x \in X \ (= G)$,

$$g \wedge x = gxg^{-1} \in G.$$

Next, if $e$ is the identity of $G$ and $x \in X \ (= G)$,

$$e \wedge x = exe^{-1}$$
$$= exe$$
$$= x.$$

Finally, for all $g, h \in G$ and $x \in X \ (= G)$,

$$(gh) \wedge x = (gh)x(gh)^{-1}$$
$$= ghxh^{-1}g^{-1}$$
$$= g(hxh^{-1})g^{-1}$$
$$= g(h \wedge x)g^{-1}$$
$$= g \wedge (h \wedge x).$$

**Solution 4.2**

(a) Let $g$ and $x$ be elements of $C_4$. Since $C_4$ is Abelian, we have
$$g^{-1}xg = g^{-1}gx = ex = x.$$

Hence any $x$ in $C_4$ is conjugate only to itself and so each conjugacy class has only one element. The class equation is
$$4 = 1 + 1 + 1 + 1.$$

*The argument here leading to the fact that each conjugacy class has only one element depends only on the fact that the group is Abelian.*

(b) As in the previous part, the group is Abelian and hence the class equation is
$$4 = 1 + 1 + 1 + 1.$$

*This part and the previous one show that non-isomorphic groups may have the same class equation.*

(c) As we saw in Section 3 (Frame 11), elements of $S_n$ are conjugate if and only if they have the same cycle type.

In $S_3$ it is easy to list all the elements by their cycle types, and the conjugacy classes are
$$\{e\}, \quad \{(23),(31),(12)\}, \quad \{(123),(132)\}.$$

Hence the class equation for $S_3$ is
$$6 = 1 + 3 + 2.$$

(d) We could list all the elements of $S_4$. However, to illustrate a more general approach, we list the distinct cycle types and decide how many there are of each type.

*The solution to this part could have been deduced directly from Frame 7A in Section 3.*

Typical elements of each cycle type are
$$e, \quad (12), \quad (123), \quad (1234), \quad (12)(34).$$

As in all groups, $e$ is in a conjugacy class of its own.

For cycles typified by $(12)$, we may choose the first entry in 4 ways and the second in 3 ways. On the other hand, the cycles $(ab)$ and $(ba)$ are the same. Hence the number of cycles of this type is
$$\frac{4 \times 3}{2} = 6.$$

For cycles typified by $(123)$, we may choose the first entry in 4 ways, the second in 3 and the third in 2. On the other hand,
$$(abc) = (bca) = (cab).$$

Hence the number of cycles of this type is
$$\frac{4 \times 3 \times 2}{3} = 8.$$

For cycles typified by $(1234)$, we may choose the first entry in 4 ways, the second in 3, the third in 2 and the last in 1. On the other hand,
$$(abcd) = (bcda) = (cdab) = (dabc).$$

Hence the number of cycles of this type is
$$\frac{4 \times 3 \times 2 \times 1}{4} = 6.$$

For elements typified by $(12)(34)$, once we have fixed the first 2-cycle, the second is determined. From our earlier calculations there are 6 choices for the first 2-cycle. However
$$(ab)(cd) = (cd)(ab)$$

because the 2-cycles are disjoint. Hence there are $6/2 = 3$ elements of this type.

The class equation of $S_4$ is therefore
$$24 = 1 + 6 + 8 + 6 + 3.$$

*We could have obtained the size of the last class by subtraction. However, doing it independently gives a check on the calculations.*

## Solution 4.3

We look at the conjugates of $e$ by all elements $g \in G$.

$$g \wedge e = geg^{-1}$$
$$= gg^{-1}$$
$$= e.$$

Hence the only conjugate of $e$ is itself. Thus

$$\text{Orb}(e) = \{e\}$$

and the result follows.

## Solution 4.4

Assume that the $s$ distinct orbits have orders $n_1, \ldots, n_s$, where we may assume that the first orbit is the one containing $e$, so that $n_1 = 1$.
Using the class equation we may write

$$9 = 1 + n_2 + \cdots + n_s$$

where each $n_i$ divides 9.

Therefore each $n_i$ can only be 1, 3 or 9.
Since $n_1 = 1$, no $n_i$ can be 9. Suppose, if possible, that all the rest are 3s, i.e.

$$n_2 = \cdots = n_s = 3.$$

Then we have

$$9 = 1 + 3(s-1).$$

The left-hand side has remainder 0 on division by 3, while the right-hand side has remainder 1.
This contradiction shows that some of $n_2, \ldots, n_s$ must be equal to 1.
If only one of these is 1, the class equation becomes

$$9 = 2 + 3(s-2),$$

which again leads to a contradiction.

Hence, there are at least three single-element conjugacy classes.

The remainders are 0 and 2 this time.

## Solution 4.5

We tackle the subgroup property first, then normality.

**Closure**  If $x, y \in Z(G)$ then, for all $g \in G$,

$$(xy)g = x(yg)$$
$$= x(gy) \quad \text{(since } y \in Z(G)\text{)}$$
$$= (xg)y$$
$$= (gx)y \quad \text{(since } x \in Z(G)\text{)}$$
$$= g(xy).$$

Hence $xy \in Z(G)$.

**Identity**  We have $eg = ge \, (= g)$ for all $g \in G$. Hence $e \in Z(G)$.

**Inverses**  Suppose $x \in Z(G)$. Then, for all $g \in G$,

$$xg = gx.$$

If we left and right multiply both sides by $x^{-1}$ we get

$$gx^{-1} = x^{-1}g.$$

Hence $x^{-1}$ commutes with every element of $G$ and so $x^{-1} \in Z(G)$.
This completes the proof that $Z(G)$ is a subgroup of $G$.

Now we deal with normality. We must show that, for each $g \in G$,
$$gZ(G)g^{-1} \subseteq Z(G).$$
Let $x \in Z(G)$ and $g \in G$. Then
$$\begin{aligned} gxg^{-1} &= xgg^{-1} \quad \text{(since } x \in Z(G)) \\ &= xe \\ &= x \in Z(G). \end{aligned}$$
Hence $Z(G)$ is normal in $G$, i.e. is a normal subgroup of $G$.

## Solution 4.6

Since $H$ is a subgroup of $Z(G)$, it is certainly a subgroup of $G$.

Suppose that $h \in H \subseteq Z(G)$ and $g \in G$. Then
$$\begin{aligned} ghg^{-1} &= hgg^{-1} \quad \text{(since } h \in Z(G)) \\ &= he \\ &= h \in H. \end{aligned}$$
Hence $H$ is normal in $G$, i.e. is a normal subgroup of $G$.

## Solution 4.7

Since the quotient group is cyclic, all its elements are powers of a generator $aZ(G)$. So each element of $G/Z(G)$, i.e. each coset of $Z(G)$, is of the form
$$(aZ(G))^r = a^r Z(G)$$
for some integer $r$.

Now let $g$ and $h$ be any two elements of $G$. Since the cosets of $Z(G)$ partition $G$, we have
$$g \in a^s Z(G) \quad \text{and} \quad h \in a^t Z(G),$$
for some integers $s$ and $t$. Hence
$$g = a^s z_1 \quad \text{and} \quad h = a^t z_2, \quad \text{for some } z_1, z_2 \in Z(G).$$
Therefore
$$\begin{aligned} gh &= a^s z_1 a^t z_2 \\ &= a^s a^t z_1 z_2 \quad \text{(since } z_1 \in Z(G)) \\ &= a^{s+t} z_1 z_2 \\ &= a^{s+t} z_2 z_1 \quad \text{(since } z_1 \in Z(G)) \\ &= a^t a^s z_2 z_1 \\ &= a^t z_2 a^s z_1 \quad \text{(since } z_2 \in Z(G)) \\ &= hg. \end{aligned}$$
Hence $gh = hg$ and $G$ is Abelian.

## Solution 4.8

Let $G$ be a group of order 9 and let $Z(G)$ be its centre.

By Exercise 4.4, $Z(G)$ has order at least 3.
So, by Lagrange's Theorem, it has order 3 or 9.

If $Z(G)$ has order 9, then $Z(G) = G$ and $G$ is Abelian.
By the Canonical Decomposition Theorem for Finitely Generated Abelian Groups, there are only two Abelian groups of order 9, namely $\mathbb{Z}_9$ and $\mathbb{Z}_3 \times \mathbb{Z}_3$.

We show, by contradiction, that $Z(G)$ cannot have order 3.
Suppose that $|Z(G)| = 3$. The quotient group $G/Z(G)$ has order $9/3 = 3$ and is therefore cyclic. By Exercise 4.7, $G$ is Abelian, and hence $Z(G) = G$, contradicting the fact that $|Z(G)| = 3$.

This contradiction completes the proof. We shall generalize this result in the next section.

## Solution 4.9

Let $G$ be a group of order 15. By Lagrange's Theorem, the order of $Z(G)$ may only be 1, 3, 5, or 15. If the order of $Z(G)$ is 15, the group is Abelian. If the order of $Z(G)$ is 1, the centre is trivial.

Suppose that the centre has order 3. Then the quotient group

$$G/Z(G)$$

has order $15/3 = 5$ and is therefore cyclic. It follows from Exercise 4.7 that $G$ is Abelian, contradicting the fact that the centre has order 3.

A similar argument shows that the order of the centre cannot be 5.

## Solution 4.10

Since we are looking for non-trivial proper normal subgroups we ignore orders 1 and 60.

Lagrange's Theorem tells us that we need only look for subgroups with orders less than or equal to 30. Hence, By Theorem 4.4 the possible orders for non-trivial proper normal subgroups are:

$$1 + 12 = 13;$$
$$1 + 15 = 16;$$
$$1 + 20 = 21;$$
$$1 + 12 + 12 = 25;$$
$$1 + 12 + 15 = 28.$$

All these are excluded by Lagrange's Theorem. This completes the proof.

## Solution 4.11

(a) This result is a direct consequence of Theorem 1.1 of *Unit IB3* which states (in the notation of that theorem) that if $ab = ca$ then $ab^n = c^n a$, for all $n \in \mathbb{Z}$.

In our case we know that $ba = a^{-1}b$, so using the above result (interchanging $a$ and $b$ and replacing $c$ by $a^{-1}$) gives that

$$ba^n = \left(a^{-1}\right)^n b = a^{-n}b, \quad \text{for all } n \in \mathbb{Z}.$$

Should you not have recalled this result you might have tackled the cases in turn as follows:

$$\begin{aligned} ba^2 &= (ba)a \\ &= (a^{-1}b)a \\ &= a^{-1}(ba) \\ &= a^{-1}(a^{-1}b) \\ &= a^{-2}b; \end{aligned}$$

$$\begin{aligned} ba^3 &= ba^2 a \\ &= a^{-2}ba \\ &= a^{-2}a^{-1}b \\ &= a^{-3}b; \end{aligned}$$

$$\begin{aligned} ba^4 &= ba^3 a \\ &= a^{-3}ba \\ &= a^{-4}b. \end{aligned}$$

(b) We start by finding the conjugates of the elements $a^i$, $i = 0, \ldots, 4$.

Firstly, conjugating by a power of $a$ leaves the element fixed (because powers of $a$ commute with $a^i$).

Next, we conjugate by $a^j b$ for $j = 0, \ldots, 4$. To do so, we use the relation $ba^j = a^{-j}b$ from part (a), together with the fact that $b^2 = e$, so $b = b^{-1}$.

$$\begin{aligned}(a^j b)a^i(a^j b)^{-1} &= (a^j b)a^i(b^{-1}a^{-j}) \\ &= (a^j b)a^i(ba^{-j}) \\ &= a^j(ba^i)ba^{-j} \\ &= a^j(a^{-i}b)ba^{-j} \quad \text{(by part (a))} \\ &= a^j a^{-i} a^{-j} \quad \text{(since } b^2 = e\text{)} \\ &= a^{-i}.\end{aligned}$$

Taken together, these two calculations show that each element of the form $a^i$ is conjugate to itself and its inverse. Hence, the corresponding conjugacy classes are

$$\{e\}, \quad \{a, a^4\}, \quad \{a^2, a^3\}.$$

Next, we consider the conjugates of $a^i b$ for $i = 0, \ldots, 4$.

We begin with the conjugates by powers of $a$.

$$\begin{aligned}a^j(a^i b)a^{-j} &= a^j a^i(ba^{-j}) \\ &= a^j a^i(a^j b) \quad \text{(by part (a))} \\ &= a^{i+2j}b.\end{aligned}$$

The exponent $i + 2j$ of $a$ now has to be calculated modulo 5, because $a^5 = e$.

However, whatever the value of $i$, as $j$ goes from 0 to 4, the expression $2j$, and hence the expression $i + 2j$, takes *all* the values from 0 to 4 (modulo 5).

Hence, the elements of the form $a^i b$ are all conjugate, producing a single conjugacy class

$$\{b, ab, a^2b, a^3b, a^4b\}.$$

The complete list of conjugacy classes is

$$\{e\}, \quad \{a, a^4\}, \quad \{a^2, a^3\}, \quad \{b, ab, a^2b, a^3b, a^4b\}$$

and the class equation for $D_5$ is

$$10 = 1 + 2 + 2 + 5.$$

(c) Since there is only one single-element conjugacy class, by Theorem 4.2 the centre of $D_5$ is trivial.

By Theorem 4.4, a normal subgroup consists of a union of conjugacy classes, one of which must be $\{e\}$. From the class equation, the only possible orders for normal subgroups are:

$$\begin{aligned}1 &= 1 \\ 1 + 2 &= 3 \quad \text{(ruled out by Lagrange's Theorem)} \\ 1 + 2 + 2 &= 5 \\ 1 + 5 &= 6 \quad \text{(ruled out by Lagrange's Theorem)} \\ 1 + 2 + 5 &= 8 \quad \text{(ruled out by Lagrange's Theorem)} \\ 1 + 2 + 2 + 5 &= 10\end{aligned}$$

Hence, there is no normal subgroup of order 2.

From the conjugacy classes, the only candidate for a normal subgroup of order 5 is

$$\{e, a, a^2, a^3, a^4\}.$$

Since this *is* a subgroup, the cyclic subgroup generated by $a$, then by Theorem 4.4 it is normal.

We might have found the subgroup $\langle a \rangle$, of order 5, and deduced that it was normal because it has index 2. However, the conjugacy class argument actually gives us more information. It shows that $D_5$ has a *unique* non-trivial proper normal subgroup which is of order 5.

**Solution 5.1**

Let the class equation be

$$p^n = 1 + n_2 + \cdots + n_s.$$

Each $n_i$ divides $p^n$ and is therefore 1 or a positive power of $p$. If each of the $n_i$, $i = 2, \ldots, s$, is a positive power of $p$ then the left-hand side is divisible by $p$ whereas the right-hand side leaves a remainder 1 on division by $p$. This contradiction shows that some of the $n_i$, other than the first, must be 1. In fact, since the number of such 1s is the order of $Z(G)$, which divides $p^n$, the number of 1s is at least $p$.

**Solution 5.2**

We apply the result of Exercise 5.1. The order of $Z(G)$ is at least $p$. Since it divides the order of the group, it is either $p$ or $p^2$.

If the order is $p^2$, then the centre is the whole of $G$ and the group is Abelian.

If we assume that the order of $Z(G)$ is $p$, then the order of the quotient group

$$G/Z(G)$$

is $p^2/p = p$. However, all groups of prime order are cyclic, so $G/Z(G)$ is cyclic, and, by Theorem 4.3, $G$ is Abelian. This is a contradiction, since the centre of an Abelian group is the whole group.

Therefore the only possibility is that the order of $Z(G)$ is $p^2$. So the centre is the whole group and $G$ is Abelian.

**Solution 5.3**

The proof is very like that in Solution 5.2, where we showed that groups of order $p^2$ are Abelian.

If $Z(G)$ has order $p^2$ then the quotient group

$$G/Z(G)$$

has order $p$ and is therefore cyclic. It follows that $G$ is Abelian and that $Z(G) = G$ has order $p^3$, a contradiction. Thus $Z(G)$ cannot have order $p^2$.

**Solution 5.4**

(a) Firstly, since $e \in H$, the coset $eN$ is in $\alpha(H) = \overline{H}$ and hence $\overline{H}$ is non-empty.

Now, let $x$ and $y$ be any two elements of $\alpha(H) = \overline{H}$. We verify that $x^{-1}y$ is in $\overline{H}$, which proves that $\overline{H}$ is a subgroup of $G/N$.

By the definition of $\alpha(H)$, there exist elements $a$ and $b$ of $H$ such that $aN = x$ and $bN = y$. Now,

$$\begin{aligned} x^{-1}y &= (aN)^{-1}(bN) \\ &= (a^{-1}N)(bN) \\ &= (a^{-1}b)N. \end{aligned}$$

But, $H$ is a subgroup of $G$ and so $a^{-1}b \in H$. Hence

$$x^{-1}y = (a^{-1}b)N \in \overline{H}.$$

This verifies that $\alpha$ is a function from the set of subgroups of $G$ which contain $N$ to the set of subgroups of the quotient group $G/N$. From the way that $\alpha$ is defined, it follows that it preserves inclusions.

(b) Firstly, we show that $\beta(\overline{H}) = H$ contains $N$ (and is thus non-empty). The coset $eN = N$ is the identity of the quotient group $G/N$ and so is an element of the subgroup $\overline{H}$. Therefore, by the definition of $\beta$, we have $e \in H$ and so $H$ is non-empty.

Furthermore, $nN = N$, for every element $n \in N$. So, by the definition of $\beta$, we have $n \in H$ for every element $n \in N$. Hence

$$N \subseteq H.$$

Now let $a$ and $b$ be elements of $H$. We shall show that $a^{-1}b$ is in $H$. By the definition of $\beta$, the cosets $aN$ and $bN$ are elements of $\overline{H}$, which is a subgroup of $G/N$. It follows that $\overline{H}$ also contains

$$(aN)^{-1}(bN) = (a^{-1}N)(bN)$$
$$= (a^{-1}b)N.$$

From the definition of $\beta$, this means that $a^{-1}b \in H$.

This verifies that $\beta$ is a function from the set of subgroups of $G/N$ to the set of subgroups of $G$ which contain $N$.

**Solution 5.5**

(a) Let $x \in \overline{H}$ and $y \in G/N$. Then,

$$x = hN, \quad \text{for some } h \in H,$$
$$y = kN, \quad \text{for some } k \in G.$$

Since $H$ is normal in $G$, and $h \in H$, we have

$$khk^{-1} \in H.$$

Hence,

$$yxy^{-1} = (kN)(hN)(kN)^{-1}$$
$$= (kN)(hN)(k^{-1}N)$$
$$= (khk^{-1})N \in \overline{H}.$$

This completes the proof that $\overline{H}$ is normal in $G/N$.

(b) Let $h \in H$ and $k \in G$.
Then $kN$ is an element of $G/N$ and, by the correspondence, $hN \in \overline{H}$.
By the normality of $\overline{H}$,

$$(kN)(hN)(kN)^{-1} \in \overline{H}.$$

Hence, $\overline{H}$ contains

$$(kN)(hN)(kN)^{-1} = (kN)(hN)(k^{-1}N)$$
$$= (khk^{-1})N.$$

Since $(khk^{-1})N \in \overline{H}$, we have $khk^{-1} \in H$.
This completes the proof that $H$ is normal in $G$.

# OBJECTIVES

After you have studied this unit, you should be able to:

(a) find all possible canonical decompositions of Abelian groups of order $n$, where $n$ is a positive integer;

(b) find the $p$-primary decomposition of an Abelian group, given its canonical decomposition;

(c) find all possible different subgroups corresponding to a divisor of the order of a finite Abelian group;

(d) find the number of permutations in $S_n$ of a given cycle type;

(e) find the conjugacy class containing a given permutation in $S_n$;

(f) distinguish between the odd and even permutations in $S_n$, and hence recognize the elements of the alternating group $A_n$;

(g) find the class equation of groups of small order, including $S_n$;

(h) use the class equation to prove results about groups of small order, in particular about the possible orders of their centres;

(i) apply the results of the unit to obtain proofs similar to those in the unit.

# INDEX

alternating group 20
Cayley's theorem 22
centre of group 29
chain of subgroups 33
class equation 27
conjugacy action 27
conjugacy class 27

conjugate 27
correspondence theorem 34
cycle form 16
cycle notation 16
cycle type 17
orbit–stabilizer theorem 26
partition equation 26

permutation group 16
$p$-group 32
$p$-primary component 11
primary decomposition 11
subgroups of finite Abelian groups
    theorem 12